D1133164

THE
BASIC CODE
OF THE
UNIVERSE

THE SCIENCE OF THE INVISIBLE IN PHYSICS, MEDICINE, AND SPIRITUALITY

MASSIMO CITRO, M.D.

Park Street Press
Rochester, Vermont • Toronto, Canada

Park Street Press
One Park Street
Rochester, Vermont 05767
www.ParkStPress.com

Text stock is SFI certified

Park Street Press is a division of Inner Traditions International

Copyright © 2011 by Massimo Citro

All rights reserved. No part of this book may be reproduced or utilized in
any form or by any means, electronic or mechanical, including photocopying,
recording, or by any information storage and retrieval system, without permission
in writing from the publisher.

Library of Congress Cataloging-in-Publication Data

Citro, Massimo.
  The basic code of the universe : the science of the invisible in physics, medicine,
and spirituality / Massimo Citro ; foreword by Ervin Laszlo.
     p. cm.
  Summary: "Explains the universal information code connecting every person,
plant, animal, and mineral and its applications in science, health care, and cosmic
unity"—Provided by publisher.
  Includes bibliographical references and index.
  ISBN 978-1-59477-391-4 (hardcover)—ISBN 978-1-59477-950-3 (ebook)
  1. Biological control systems. 2. Homeostasis. 3. Life—Origin. 4. Matter.
5. Cosmology. I. Title.
  QH508.C58 2011
  500—dc23

                   2011020697

Printed and bound in the United States by Lake Book Manufacturing
The text stock is SFI certified. The Sustainable Forestry Initiative® program
promotes sustainable forest management.

10 9 8 7 6 5 4 3 2 1

Text design and layout by Priscilla Baker
This book was typeset in Garamond Premier Pro with Helios and Gill Sans used
as display typefaces

This book is dedicated to my teachers—

Giovanni Lever

Bruno Bruni

Siro Rosseti

Alberto Sorti

Adele Rosso

Pepe Alborghetti

Hans Christian Seemann

Remo Bessero Belti

Dino Sartori

Paulo Parra

Lino Graziano Grandi

Alessandra Zerbinati

Francesco Aramu

Anna Valente

Claudio Cardella

Gabriele Mandel

Nirmala Lall

Fausto Lanfranco

Angelo Pippo

Agnese Cremaschi

Adele Molitierno

Emilio Citro—

And to all the people who taught me something.

CONTENTS

o—●

FOREWORD
Ervin Laszlo

The present book by Massimo Citro is actually two books in one. It is the presentation of an original piece of research of great practical interest, and it is at the same time the outline of a new scientific paradigm. In both respects the book is detailed and remarkably complete; hardly anything remains to be added. Thus I can confine myself to some brief observations, and above all to a recommendation: read this book and read it carefully, and note the key assertions. They are basic points of reference for anyone interested in a revolutionary way to obtain the benefits of natural substances and medications in the "pure" form where the information that defines their properties is transferred without transferring any molecules—in this regard this is a variety of homeopathy but achieved with a different technique. They are also, and from my point of view most importantly, basic points of reference in the world picture that is presently emerging at the frontiers of the natural sciences.

The fundamental premise of both these presentations is the same, and it is that that enables Citro to present them as organic parts of a single book: it is information. It is information not just in the mind but also in nature, information that is omnipresent and ever determinant, yet invisible to the bodily senses. This information is nonetheless regarded by the currently dominant scientific paradigm as being of questionable relevance and reality. In this regard the dominant scientific paradigm is in urgent need of updating.

According to Citro, information is the "other side" of things. I agree, but I would go even further. Information is not merely the other side of things, but the fundamental element of the cosmos: it is the element that constitutes its basic reality. The manifest, sensible reality is but the result of the action of information on what Citro, following Newton, calls *material pura* (pure matter). The universe as a field of information is the new concept of reality, more exactly, the newly rediscovered perennial concept that has been ignored and even denied in the mainstream of modern science.

In the view I have been elaborating for the past two decades—stated most clearly in two recent books, *Science and the Akashic Field*[1] and *Quantum Shift in the Global Brain: How the New Scientific Reality Can Change Us and Our World*[2]—the fundamental furnishing of the universe is information and energy, rather than matter in space and time. Information is the "software" that governs the mass/energy "hardware." What differentiates this universe from any other actual or potential universe is the information it contains. The laws of nature are, in the last count, information-based algorithms governing how manifest entities in space and time act, react, and coevolve. As Citro also points out, the reason that things conserve their characteristics and identity in space and time is due in the final analysis to the consistency and persistence of the information that defines how they act, interact, and evolve—and thus, in a realistic (if epistemologically naïve) context, what they are.

Information, evidently, does not exist without a physical basis: that basis is what endows information with eminent reality. Information is present in nature in the form of a basic code and a physically real field: Citro's "informed field" and my "Akashic field." This is not an ad hoc postulate, a field suspended in the limbo of mysterious phenomena, but a fundamental element of the universe. It is not the zero-point field of the quantum vacuum, but a cosmic plenum. In view of its paramount reality as both the origin of the energies that bundle into the packets we know as matter (at the presumed beginnings of this universe) and their ultimate destination (in the final "evaporation" of black holes), and in view of its capacity to conserve and convey information, the cosmic plenum recalls the ancient concept of Akasha. It is properly called "Akashic field" (or A-field for short).

Theories as to how this field relates to the domain of manifest entities are as yet hypothetical: here we must bid for patience—work is in progress. Most likely the interaction dynamics of the A-field of the plenum and the entities of the manifest domain of space and time involve scalar rather than electromagnetic waves, and holographic interfering wave-fronts rather than linear interactions. In a hologram information is in a distributed form (i.e., it is present at all points); as a result, the interaction of the A-field (or informed field) with the entities of the manifest domain proceeds not merely through classical "upward" causation, where parts influence the wholes formed by them, but also through the inverse "downward" causality, where systems exercise an influence on their parts. Through holographic downward causation the entire universe, a field-connected integral system, acts on all its parts: on every particle, atom, molecule, crystal, cell, organism, and society or ecology of organisms. This influence is subtle but real, and needs to be recognized by scientists, just as psychics, artists, spiritual leaders, and visionary philosophers throughout the ages have recognized it.

Let me conclude by repeating the recommendation with which I started: this "two books in one" contains important and in some respects essential knowledge. On the one hand it demonstrates that the neglected "submerged" side of things can be experimentally verified in ways that can make a significant contribution to our health and well-being. And on other it provides a cogent exposition of the new scientific paradigm that can account for the observed facts, a paradigm that changes our most basic understanding of the fundamental nature of the universe and of all things in it.

Ervin Laszlo, twice nominated for the Nobel Peace Prize, is editor of the international periodical *World Futures: The Journal of General Evolution* and chancellor designate of the newly formed GlobalShift University. He is the founder and president of the international think tanks the Club of Budapest and the General Evolution Research Group and the author of eighty-three books translated into twenty-one languages. He lives in Italy.

PROLOGUE ⦿ THE OTHER SIDE OF THINGS

Jeremiah Johansohn was only minutes away from his death when he finally understood what had happened to humans after our expulsion by God from the Garden of Eden. His insight came from the method of enquiry that, like a law, he had observed all his life. He called it "the game of the riverbank." He had first sensed this game long before, when the purpose of life had seemed clear, even if only as a reflection. It was the day of new clothes, perhaps for communion, when in the afternoon, deliriously happy at the scoring of a goal in the children's practice match, he had embraced a chestnut tree in the garden of his grandparents. While hugging it, he had felt a vibration conveyed through the trunk, and looking up, he had seen the fine ends of the branches moving, as if brushed by the wind, but there was no wind there. He did not run away, but this first encounter left a mark in his memory and his soul.

In the years that followed, he focused his interests within the fields of science and biology, and by the time of his graduation, he was well acquainted with the world of pharmacology. The burning question for him was not so much what human beings are or what our fate might be, but rather why? Why are things natural and humans not? Why are animals natural and humans not? Why is creation so perfect and humans not? It was in response to the last question that Jeremiah wrote in his laboratory book in clear, forceful handwriting, "Why has humanity missed something, something that the creator sees but we do not? Perhaps the expulsion from paradise was really about losing the vision of the greater part of things? Each element of creation lives in an eternal state of consciousness and is intrinsically finished and complete. In that realm of unknown things death does not exist. With that vision, there is no need for enquiry, as with a movie seen many times. It is only humans that are blind within creation, humans that call out from every angle for an explanation of the film."

Over the years, documented in his notes, Jeremiah came to understand more and more of what humans miss. Dogs sense the approach of their masters; whales pass through the exact same points of longitude and latitude on the globe year after year, missing their appointments by only a few days; plants and trees have emotions similar to ours.

"See," he said one day to one of his young assistants, "we also have to follow rules and must enter a precise code in order to express the vision of the universe. That code is made up of many tiny codes, which are behavioral formulas. Creation is born and grows and then dies following precise formulas. It continues to recreate because the codes, the formulas, are not in the creations themselves but are rather outside of them. Remember: outside of them. We have no proof yet, in the form of exact deductions, but years ago a tree gave me back the same happiness I gave to it and I was no more crazy or a visionary then than I am now. If a tree can be stirred by a different life form then the whole universe is speaking the same language. Take a drug for example. Is there a way in which the actions of medicines can be obtained without actually administering them?"

He looked at the expression on his young assistant's face and laughed. "I understand that you are now worried for my safety: if this were known, think how the multinational pharmaceutical companies would look for me. I would be at great risk of being fragmented into a hundred thousand tablets or bottled up as a syrup."

He laughed again before frowning and continuing, "Each drug, when acting properly, is like a key in a lock. It enters into a receptor and triggers a mechanism within the cell. But it does more than that. You can imagine it as a sound if you wish and its action sounds like a song. In reality it is a coded frequency that reaches the 'ears' of the whole organism. The beauty is that the organism recognizes the song and acts accordingly, a bit like the way in which music can affect your life. I discovered, however, that it is not necessary to put the whole orchestra inside the human body. If you can put the specific tune there that cures a headache, an ulcer, or colitis, then you no longer need to swallow saxophones, pianos, oboes, drums, harpsichords, trumpets, and violins, all of which are largely indigestible. Each medicine emits specific vibrations and coded effects as electromagnetic waves, definable wave frequencies. If I eat a piece of pear with the same 'music' as an antineuralic, the body hears the echo of the original drug and the cells are stimulated to behave and react just as they would by the action of the actual medicine. This is similar to recording a live concert and then listening to it quietly in the comfort of your own home."

The assistant looked at the bearded face of Dr. Johansohn in astonishment before pretending to smile as Dr. Johansohn explained, "Boy, this is all true, but that is not to say you can avoid dying. Death cannot kill a walking pharmacy, but it can kill a person who has reached the limits of possible maintenance.

"Drugs are like tiny radio stations emitting their own information that summarizes their identity; it is a kind of signature, a digital fingerprint, like a bar code." He paused for a moment as if remembering an appointment. He walked a few steps away, then turned to his assistant as if to ask for confirmation: "If this is true for medicines, might it not be the case that all matter releases its own signature, like an invisible web of communications spanning the entire universe?"

The assistant adjusted his white coat as a sign of both surprise and excitement. He had followed the speech perfectly, and it was not simply out of respect that he had no answer to offer, but rather another question.

"And if this is the case . . . ?"

Jeremiah Johansohn stared at the wall behind the young aide as if he were reading the answer on the white tiles of the laboratory. Then he adjusted his focus and returned to look at his interlocutor: "If it is so, and I have no doubt that it is, then dear George, welcome to the other side of things."

In the years to come, after the death of Dr. Johansohn, George collected together the many papers that scattered the floor of their laboratory. But he didn't stop there. He went on to decipher the experiments, to write essays, and to put together the book his teacher had wanted to publish, but didn't, in order to avoid disappearing into a bottle of syrup at the hands of the multinationals. He had thousands of copies printed and surreptitiously left the books at places from Philadelphia to Boston, from Rome to Paris.

"What?" asked the waitress in a diner in Rochester, Vermont.

"A treatise on physics. Or rather, a philosophical journey, perhaps a hard journey but also easy, as easy as turning a big sock inside out."

"A big sock—how big?" she asked with her face betraying more interest in other things.

"As big as the universe" responded George, before sinking his teeth into the basic code of a cheeseburger . . .

INTRODUCTION

I f you dream of understanding the reasons why things happen and how the world is made, if you are fascinated by the stars and elementary particles, if deep curiosity about research and an urge to make progress direct your life, then this book is for you and I will tell you a story. It is the story of a journey into the other side of things.

Do you remember your first questions: who are we, where do we come from, where do we go, is the universe finite or infinite, what is eternity, is there life on other planets, what is the atom made of, and so on? How many of these questions have really been answered? This book does not propose solutions, but rather suggests avenues of research and points of view you can use to trace knowledge along an ancient path that is ever new.

In my school days, I wondered what marked one body as different from other bodies, since they are all basically made from the same ingredients. As all matter is formed from only ninety-two elements (and a few others that at that time were still to be discovered), how can there be such variety? In addition, these ninety-two elements are in turn composed of particles that are always the same. Think about it: the elementary particles are the same throughout the whole universe. Our brain has the same cloud of electrons as a stone, a leaf, and a whale. The more we probe matter under the microscope, the more it appears amorphous, formless, without those very differences that distinguish bodies and substances. If oceans, mountains, plants, animals, and all other things are aggregates of elementary particles, what gives them their material shape? Different things are differentiated by their form, not by the particles that actually make up the matter of which they are made. What is it then that determines identity? Does form give meaning to matter?

Aristotle writes of a "basic matter" or "prime matter" (ὕλη ἀμορφή, matter without form), the origin or basic matrix from which all bodies arise. What relationship can there be between the Aristotelian model and the amorphous material decomposed into elementary particles? The key to knowledge is not limited to particle physics: research keeps on finding new and even smaller particles, but does not reveal the secrets of existence or tell us what things are ultimately made of. The reason is simple: research insists on investigating "from our side."

The first requisite in order to obtain access to "the other side" is to free ourselves from the concept of necessity, on which our investigative mechanisms have so far been based. Ever since the expulsion from the Garden of Eden, evolution has always been prompted by necessity, from which no living being can escape! Necessity was then Ananke, the Greek mythical divinity. Events were seen as having no causality, instead happening because they were necessary. Even the meaning of things was thought to always be born of the goddess Ananke, by necessity. If matter were invisible it would not matter! It becomes of consequence only when registered by our senses, and this in turn occurs only when we deem it necessary. No one is interested in what happens on one of Saturn's moons! Physical bodies have no meaning to us without our interest in them; to arouse our interest, they must be considered relevant on some level. An item of showroom furniture has no meaning until we start to consider buying it and how it might look in our home.

Knowing comes from the need to survive and from curiosity. Behind both lies fear. Each form is determined by function, the function of need and the necessity of fear, without which there would be no evolution. Knowledge is also a medicine for fear and in the past stemmed from seeing, seeing with the eyes and also with the mind. In ancient Greek, the verbs *to see* and *to know* have a common root, *vid:* "to see with the eyes," "to see with the mind." Today, fear still allows us to know but not to see. Humans have tried to remove as much fear as possible from within the twenty-four hours, reducing it to a few minutes a day. But nature invents needs in order to organize programs.

Today scientific research continues to investigate the part of the world to which we give meaning. If we want to gain access from the other side, we must distinguish between the meaning of things and the meaning we attach to them. On the one hand is the meaning that the universe gives to itself and on the other the meaning that we allocate to it, and each one does not care about the other. Imagine if suddenly the universe decided to disobey its own laws, its own meaning, pausing its activities for just a minute: everything would collapse, including us! Fortunately the universe is not a member of a trade union, and it doesn't go on strike. We cannot even imagine the sheer scale of it, nor have we any idea of how big it may be or the form it might have. If it is formless to us and beyond our comprehension, then it loses meaning and becomes of no interest. Out of seven billion humans, how many are in fact scientific researchers dedicated to the study of the universe? Probably less than the number of flies that hit the windshield of a speeding car.

It is form that gives significance to matter. The sand in a bucket tells nothing to the little boy who plays with it; it lacks meaning. The woman who is separated from her husband by the expanse of the Namibian desert might see that desert, that sand, as the pure expression of long-sought freedom. But the same husband, lost in that desert, will give it a very different meaning, and he may even die of thirst because of it. Yet it is the same sand that their little boy plays with in his bucket.

The big bang—if indeed it occurred—was an explosion from which sprang our heavenly bodies and the laws that still seem to be governing the expansion of the universe. Out of this infinite array of possibilities, we seem capable only of attaching significance to the things we need. Research on this basis will always be misleading and incomplete.

It is necessary to explore the other side in order not to stand still, just counting particles and talking about them. Such exploration is what many great people in history have done—Pythagoras, Plato, Leonardo,

Bruno, Newton, Leibnitz, Mozart, Einstein, Böhm, some other quantum physicists, Eastern mystics, Western poets, teachers, saints, sages, and philosophers—when physics was not yet divorced from philosophy! They have left traces everywhere of one big secret that has been handed down through the centuries, both concealed and revealed. Each in his or her own way, appropriate to his or her own time and culture, has told us a story, a story at times scattered diversely through forests of symbols and enigmas. But if we look again closely, using the keys provided in this book, we might yet still see a magical thread leading us to the very origin of the biggest secret in world history: *the proof that the world in which we find ourselves is a perfect, immense, deceptive, virtual reality, made of nonexisting matter.*

Have you ever wondered how nature creates itself? Do you agree that, before making something, you should draw up a plan that includes the information necessary to build the forms and architecture, parameters and relationships, materials and facilities? Any construction, from a toy to a skyscraper, must have a design beforehand. So does Mother Nature.

The universe is a design that evolves through becoming. The word *nature* derives from the future participle of the Latin verb *nascor,* "to be born"; it indicates that the essence of a thing is not what it seems, but *what it is not yet* and will be: the design of its future projection. Consequently, physics, (from φύσις, "nature") has to investigate not only what is manifested (quantifiable and measurable) but also and especially what is *not yet:* the design that holds the world together.

I began to deduce the secret of matter during my years spent in the laboratory. I discovered that the universe is full of codes that define and inform the nature of things. *To inform* means, first of all, "to provide form," and these codes appear to play important roles in the architecture of bodies: their structure, characteristics, quality, and functions. They also regulate growth and development. You can think of these codes as being something like bar codes or fingerprints.

I want to share with you the fascinating things I have learned about these codes—which I term *basic codes*—that operate at every level of reality, functioning as a matrix, a regulating system, and a means of communication throughout the universe. A few fundamentals, which we will explore in detail, are:

- The basic code is a set of essential data that define the field of a substance, then the form. Thanks to this information, the code acts as a first draft, the map from which the body derives its structural references. The code governs characteristics such as spatial extension and borders.
- The basic code regulates the homeostasis of the body, guarding its form, unity, features, and functions. In cellular organisms, it plays the role of an intrinsic control system.
- The basic code gives matter a rhythm, making the space around it vibrate; information in the form of rhythmical sequences radiates to the field, which thus remains informed.
- The information field allows the body to communicate its being and characteristics to others through field interactions.

In order to share my discoveries with you, I invite you to take a journey with me, a journey into the essence of matter and beyond, into the mysteries of the field of information pervading and connecting everything. Our research into the design of the universe requires a new method, a newborn physics that does not yet have adequate means to confirm hypotheses, for there are only clues of the existence of codes (whose nature is not molecular) to help direct this pursuit. We will explore where there is no perception, the emptiness that is not empty. The other side of things.

1 ⊶ PRELUDE TO MATTER

We Live by Perceptions

We perceive the environment through our senses. The body relates to the external world via various kinds of stimuli that excite the retina, the cochlea, the acoustic nerve, and the myriad of nerve endings for touch, taste, and smell. These receptors transmit signals to certain areas of the cerebral cortex, where they are decoded and translated into visual images and other sensory sensations. If we consider the nature of the stimuli, there are three senses: sight, hearing, and contact (taste, smell, and touch); sight is activated by electromagnetic waves, hearing by elastic waves,* and the other three by direct contact. Our *sensations,* as the word implies, are not certainties but "senses" of something.

Further, our sensations are subjective by nature; the same stimuli can produce different sensations in different brains: it is the brain then that transforms. Let's see how. Humans generally believe that everything that happens is grasped by the senses, but it is not so. Our neurons can process only a fraction of the signals we receive from the environment,

---

Elastic waves are mechanical waves that propagate on the surface of a medium, without causing permanent deformation of the medium. *Encyclopedia Britannica* adds, "If a material has the property of elasticity and the particles in a certain region are set in vibratory motion, an elastic wave will be propagated. For example, a gas is an elastic medium (if it is compressed and the pressure is then released, it will regain its former volume), and sound is transmitted through a gas as an elastic wave."

as if there is a screen up between us and nature that only allows through some light, sound, and tactile frequencies, which are then translated into *mental images*. What we see becomes images; hearing produces images, as does everything we touch, breathe, and taste. What we think: images. What we dream: images.

The universe for us is only what the brain can obtain and understand by translating sensations into mental images, from which framework we derive learning and knowledge. Reality and imagination are translated from mental images. What we take from outside or what we produce from within is always *vid*—seeing with the eyes and mind. We live by sensations and therefore by images, in a triangular game played between rationality, imagination, and necessity, where the dosage of need determines the most appropriate rate of exchange between rationality and imagination.

To the incompleteness of our senses we need to add the subjectivity of our processing. Our mental images are primarily species specific in the sense that the world we see is not the same as a dog sees, or a lizard, or an insect, because each species has different neural configurations responsible for mental representations. There can be diversity even within the same species: just look at the changes produced by the brain of a color-blind person. The apple that is red to me and green for a color-blind person, in reality has no color. Even the colors belong to our "internal cinematographic system" and are formed in a certain way because the neurons are in *that particular way*. According to the same principle, it is possible that not even the shape of an apple is as we perceive it. Yes, we can touch it; it has a texture, a smell, and a taste! But these attributes are only the results of subjective neuronal responses. This is the way our limited and subjective senses paint the world for us.

Reality remains veiled; nobody knows it, and everything is an interpretation. We are "blind to the world"; we are not looking outside, but rather *inside* our own heads. We are reading from the brain, watching a film that runs on our cortex. We are prisoners of an inner world, of a machine that produces a virtual reality. So it ends that the senses— our only means of contact with the external world—keep us separated from it through representations that are not real. Sensory data are misleading, like in the Indian allegory about blind children confronting an elephant for the first time. As each child touches a different part of the elephant, each creates a different mental image, each arrives at a different interpretation. Although most of the material world is not visible to the senses, we presumptuously think to confine the entire universe to our perceptions—"I don't perceive it, so it doesn't exist"—as if the world was modeled on us!

The fact remains that we perceive only a negligible portion of the vibrating ocean in which we are immersed. We fail to detect the infrared and the ultraviolet, infrasound and ultrasound, and in general the very high and very low frequencies; we can't even detect the X rays, gamma rays, radioactivity, and cosmic rays, which all still affect our bodies. And so many frequencies are still unknown. The senses are therefore incomplete; our neural circuits can't process the majority of inputs in order to translate them into images. According to some,[1] our senses comprehend only 5 percent of the signals from the world, which means that we miss 95 percent of our environment. In modern physics dark (imperceptible) matter is inferred to account for 23 percent of the mass-energy density, while ordinary matter accounts for only 4.6 percent, and the remainder is attributed to dark energy. These figures indicate that dark matter constitutes 80 percent of the matter in the universe, while ordinary matter makes up only 20 percent.

Even though the senses are not able to detect most of the universe, the invisible nevertheless exists. As the Jungian Dr. James Hillman points out, we live surrounded by crowds of invisibles, such as time, ideals, values, abstracts, and all "people" who were deified in ancient times. By *invisible,* I mean "not visible," that which cannot be seen, or touched, or detected by any of the senses. Humans have talked with the spirits of things and been visited by the gods; the invisible once had a place in the everyday, like the angels above Berlin in the film by Wim Wenders. In the course of time, the invisible was confined to the categories of the fantastic and mythic, while science has become increasingly the "logic of solids," to quote Bergson. Logic no longer distinguishes between magic, mysticism, and the scary fantasy of monsters. Nowadays, mythical thinking is crushed by the aridity of reasoning, and the invisibles are dismissed, causing a loss of idealism, feeling, inspiration, intuition, and creativity. Apart from the poets, humanity has become blind and incapable of grasping the invisible essence of nature.

To the invisible, it matters little whether it is perceived or not: like a fixed star, it follows its own path. We are not able to see air, for example. Yet it exists and permeates everywhere in the world. Forbidding us to see certain things, Mother Nature allows us to see others, without which it would be impossible to exist—the dictates of necessity once again. The problem is that we strive to reach an absurd science, established purely on objective methods, reproducible and absolute, when in truth scientific research is based on sensory perceptions and our senses are quintessentially limited and subjective. The parameters used for the visible are unable to prove the invisible: we may feel the presence of the invisible without being able to confirm it. We must then accept the idea that a research into design proceeds in large part in the invisible, far from the logic of solids. We begin with matter itself.

To Encounter Pure Matter

Quantitas materiae est mensura eiusdem orta ex volumine ac densitate conjunctim: "The quantity of matter (of a body) can be accounted for and measured from its volume together with its density." So begins the most important work of Isaac Newton, *Philosophiae Naturalis Principia Mathematica,* his treatise on gravitation.[2] Our journey begins with him in England in 1687, when he published the work that made him famous, the father of modern science, authoritative above all suspicion. But Newton was also an alchemist, and he knew of many things.

From his very first words he points out the difference between matter and mass. Why? Are not mass and matter synonyms? He adds, "And from now on, I intend to designate the quantity when I use the terms *body* or *mass.* The same (quantity of matter) is known by the weight of each body, which is proportional to its mass." In other words, the mass of something is matter combined to form a body, gaining density and volume. But beware: if *combined matter* attains weight and extension in space, then does matter that does not combine itself lack these properties? Here is the beginning of a problem: it is not easy to think of matter that has no volume or density, that is "without form," so to speak. There is no place in our categories of thought for something without boundaries: thoughts are made of images, and images are forms with distinct boundaries, like masses that express the "quantity of matter" in an aggregate body.

But, in fact, matter is without form; "it is amorphous," as Aristotle described *prime matter*. Newton was aware of this matter, which eludes the senses. He knew of it because it was spoken of within the alchemical tradition. He understood that we have to deal with two parallel dimensions: one of matter (which we cannot perceive) and the other of mass (which we can perceive). Matter is not visible, nor can it be represented, but we can realize it with our imagination and intuition. Great scientists always use their imagination. Intuition is like a truffle dog that the scientific philosopher sends into the night to find the prize, but once the location is found then there will be years of digging ahead.

Three centuries after Newton, the idea of prime matter was revived by the well-known Italian mathematician Francesco Severi from Arezzo. In 1947 he called pure matter (*materia pura*) "not differentiated" (formless, without quality), not subject to time (that is, motionless and in perfect stillness with regard to any observer). This is matter with no mass (having zero rest mass).[3] Matter free from time is eternal. Invisible, intangible, timeless, eternal: how can the mind imagine something like this? What does it mean to say that the matter is *pure*? It is not about moral purity; rather *pure* means that it is not yet combined into a body, not determined by any bodily form, having no specific identity, but holding everything in potential. Now do you realize why pure matter escapes our senses? It does not become and it cannot become because becoming is transformation, and pure matter cannot be transformed. It is not exactly nothing, but it is *prope nihil,* "close to nothing," to quote Saint Augustine.

Severi's ideas were shared by Francesco Pannaria and then by Claudio Cardella (forming the Physics of Severi-Pannaria-Cardella). Pannaria was one of the greatest chemists and physicists of twentieth-century Italy, the last witness of the so-called boys of Via Panisperna.* Inspired by Severi's physics, Pannaria argues that the universe is made up of a "world stage," which includes everything perceived by the senses, and a "backstage," which is made up of pure matter that we cannot perceive. The theater's backstage is the matrix from which detached forms appear on the scene. Pannaria writes, "The backstage of the physical world, the anti-world of our world, is the vast sea (*mare magnum*) of pure matter, the canvas on which is embroidered the history of the universe."[4]

Bodies are formed when pure matter is combined and in so doing loses its purity. From being uninterrupted and continuous, it fragments in bodies, in alternations, and in events. It becomes a rhythm. Time and space, which were previously absent, appear on the scene, and from that moment on matter has mass. A look at nature can help us to understand how matter can change from formless to the formed, such as when the lava of volcanoes frightened our ancestors, yet lost its mystery upon turning cool and becoming rock. The same applies to metals and water. The heat we identify with fire is one of the means by which matter can be combined to manifest as mass with form. Matter has no form but mass does: we can perceive and relate to it. Our mind can understand it and remember it. We do not remember stones encountered on a walk, nor does the miner recall the rocks he has extracted, because their forms do not represent anything meaningful.

---

*According to Wikipedia, the boys of Via Panisperna (*I ragazzi di Via Panisperna*) were a group of young scientists led by Enrico Fermi. In Rome in 1934, they made the famous discovery of slow neutrons, which later made possible the nuclear reactor and then the construction of the first atomic bomb. The nickname of the group comes from the address of the Physics Institute at the University of Rome La Sapienza, on the street named Via Panisperna.

According to Aristotle, all the nonexpressed elements are in pure matter, principally the four prime ones that, according to Empedocles, formed the physical universe: earth, water, air, and fire. Pannaria also speaks of the four elements, which in modernity are referred to as matter, mass, energy, and field. We have already looked at the difference between matter and mass; now let us see how they correspond to energy and field.

Enlightened Energy

Einstein's mass-energy equation says that $E = mc^2$, which means that at the square of the speed of light, mass tends to transform into energy. The consequence of this is the wave-particle duality. This is supported by quantum mechanics, which states that particles, atoms, and molecules obey mechanical laws different from those regulating bodies of larger dimensions. It is based on the theory formulated in 1900 by Max Planck, indicating that energy has a discontinuous structure, consisting of discrete packets (quanta), and that light thus consists of photons and particles. Quantum mechanics has since demonstrated that waves and particles are different aspects of the same thing; mass and energy are interchangeable. At very high speed, mass turns into energy and energy can also become mass. This effect has been observed in bubble chambers used for nuclear experiments, when the passing of high-energy electromagnetic waves through heavy atomic nuclei results in a transformation into particles and antiparticles. Photographs show that the energy converts into mass and the wave becomes a body.

We will see that this perspective relating to masses converting into energy and vice versa is not strictly correct, because in reality the mass is not mass, but rather energy distorted by the senses, by perception, an interpretation *ad usum sensorum* of events that occur in a different way from how we register them. But to keep things simple for now, we will continue to speak of mass and energy. What is important is to realize that this world is an immense virtual reality—a dream, an illusion—where everything and even the opposite of everything is possible. But let us continue step by step.

In quantum physics, wave and particle coexist alternatively, in the sense that when it is not observed, energy appears as a waveform, but when we observe it in any way, energy "becomes" particles. According to the uncertainty principle enunciated by Heisenberg, we can't grasp reality in its entirety, but can only witness one of its possible aspects at a time. His principle indicates that in the subatomic world it is not possible to determine for sure the position of a particle and its speed at the same time. The more precisely we get to know a particle's position in a given moment, the more uncertain is the determination of its speed, and vice versa. So we are limited to the determination of a singular function and remain in darkness concerning the other. The principle of indetermination is also expressed in the relationships between other dimensions, such as the interval of a reaction and the energy involved: events that happen in a brief instant involve uncertainty of energy, and vice versa. No transformation is taking place. Rather we are interpreting the manifestation of microscopic entities in two different ways, sometimes as energy and sometimes as corpuscular forms. Understood in this way, mass is energy that the senses interpret as mass. Therefore, particles would not even exist if it weren't for the capable illusionists called senses. All is illusion. The senses describe the world not as it is, but as they understand it. Particles and waves are the same thing regarding the alternative functions of mass and energy. What differentiates them? Mass is defined and can be measured and weighed. Energy cannot.

Mass has extension, weight, speed; it is made of molecules and atoms. But what exactly are atoms? We owe the idea of the atom to Democritus of Abdera (ca. 460 B.C.E.–ca. 370 B.C.E.), the head of the Greek philosophical school called atomism. He stated that the world is formed by myriad indivisible particles, atoms (*atom* literally means "nondivisible"). If they were infinitely divisible then they would dissolve into emptiness. Democritus agreed with Parmenides and the Eleatics that "only being is."* According to Parmenides, "Being has never been, and never will be, because it is now a whole, one and continuous. . . . Nor is it divisible, since it is all alike; nor is there any more or less of it in one place which might prevent it from holding together, but all is full of what is."† Democritus also identified being with *full* and nonbeing with *void*. Full of what? Countless tiny invisible elements that can't be divided further. The atomists regard them as the solid, extreme frontiers of the world stage. An atom is the smallest conceivable "thing": it can be weighed, measured, and most of all, perceived.

Years ago, when my generation went to school, the atom was taught according to physicist Nils Bohr's model: particles in the shape of balls, depicted like planets of a tiny solar system orbiting around the nucleus. It was an incomplete and debatable model, but it was clear and tangible. At the same time we were also taught that the "balls" were not in fact solid, but made of electric charges. And here, superhuman efforts were required to accept or even to imagine that it was exactly those electrical "charges" (and God only knows how the human mind can imagine them!) of impalpable nature that were the bricks that make up the solid and heavy bodies of a physical universe.‡

Then everything changed. Some people started to speak of "energy quanta," of "electronic clouds," describing electrons as "probability

---

*Parmenides was born in Elea, a Italic colony in ancient Greece, in the sixth century B.C.E. Said to have been a student of Xenophanes of Colophon, he founded the Eleatic philosophic school. He exhorted man to go beyond knowledge gained by the senses and to use reasoning to explore the world that is not influenced by senses.

†This is a complex and profound concept of *being;* being is necessary and real, while nonbeing is possible but illusory. According to Parmenides, "It is necessary to speak and to think what is; for being is, but nothing is not." Being is similar to Pannaria's pure matter because it can't become, otherwise it would be determined and thus would become nonbeing. Platonic thinking is based, in large part, on Parmenides' thought. In Platonic thought nonbeing becomes the world of *doxa*, "opinion or illusion," while being is *aletheia,* "the truth, the real world, the other side of things."

‡For example, when I think of salt, I can imagine a sodium chloride molecule, which I can associate with a distinct flavor, taste, color, and sensation to the touch; I can imagine sodium and chloride by knowing their properties. But beyond that, it is difficult to conjure up an image of a proton or an electron!

waves" or "spheres where electrons are probable." At this point the mind was lost because these things cannot be envisioned. A "wave of probability" is not a readily available image. We can't present an image of the ἀρχή (archè), "the principle from which all originates." As in the case of the *apeiron aòriston* (undetermined infinite) of Anaximander, it is not understandable because our brain needs specific references.*

In 1984, the American physicists Michael Green and John Schwarz suggested that the *superstrings theory* could explain the nature of matter. But be wary of the explanation, as it is perhaps too easily "music to our ears." Formulation of the string theory, in which several physicists were involved, began in the late 1960s when a young theoretical physicist, Gabriele Veneziano, deduced (using a formula of the great eighteenth-century mathematician Eulero) that particles were not actually formed as points and in fact were not even particles at all, but rather thin strands (like microscopic rubber bands)—in fact, strings—that vibrate constantly. These strings would be so small so as to appear like points. Their dimension would be around the "Planck length," in other words a hundred billion billion (10^{20}) times smaller than an atomic nucleus. According to the theory these strings would be the smallest constituents of matter, making up particles and atoms. This theory attempts to be compatible with Einstein's theory of general relativity (according to which the observed gravitational attraction between masses results from their warping of space and time) and with quantum mechanics. The superstrings theory says that the variety of masses and elements depends

---

*Anaximander of Miletus, one of the first pre-Socratic philosophers, states that the origin and cause of the whole universe is infinite, undefined, pure quantitative and extended matter, unlimited, divine, but most of all, undetermined, because the elements are not yet distinct. Out of this infinite extension different ingredients mix together and separate out to be identified as elements. This concept is another forerunner of that of pure matter. The undetermined infinite is a reality separate from the world, transcendent, but it also is the law of the world. It is an eternally moving amalgam out of which the elements, the bodies, and the universes emerge, necessarily separated; from this separation starts the opposition of contraries. There are many universes that all derive from the unlimited that embraces and governs them, but without will and personality, and to which they will ultimately return and be dissolved.

only on the different ways in which the strings vibrate. Keep this in mind, for later we shall return to this musical theme of vibration, and to matter as eternal music . . .

Although they exist, ultramicroscopic subatomic structures of any kind cannot be represented. Heisenberg teaches us that in the microscopic world, we have to change our categories of thoughts. There we find no dimension, weight, form, size, color, taste, or anything else from our macro world. Each world has its own laws that are inviolable, and communication is possible only if we learn the language, time, and space of that world. When we relinquish the idea of solid bodies, new concepts that the mind can hardly imagine emerge, such as energy in the form of light, fields as contours bordering empty space, waves and pulses with rhythmic movements. The physics of the invisible, which explores the other side of things, is essentially the physics of the fields.

The Field, the Reality of Things

Mechanistic thought conceptualized solid particles moving in a vacuum. Then came field physics, and prevailing notions were shattered once again. In the mid-nineteenth century, Michael Faraday introduced the idea of a *field* as "a space around a source of electromagnetic energy." Opposing the concept of "full and void" from atomism, Faraday suggested the idea of "matter and force diffused in space," according to precise lines of force. His was a nonmaterial vision of physical phenomena! It is with Faraday that fields became defined as physical dimensions in zones of temporal space. In the following century, Einstein extended the field principle with the inclusion of gravity: the universe is thus considered held in a single gravitational field that curves in proximity to matter.

Of the four elements of Pannaria, the field is the least studied but the most interesting. Mass could be matter combined with energy, which is an expression of the field. In that case mass would be the formation through which the senses perceive the field, the reality that the "veil of Maya" hides, as some insightful sages of India, along with some

Western philosophers, have put it. Plato contrasted the truth (*alètheia*) with fiction, opinion, illusion (*doxa*). The senses fall under the category of doxa, projection, the shadow of the alètheia. The senses enable us to perceive only impressions, while the truth of the universe is unknowable. "Nature loves to hide" (Φύσις κρύπτεσθαι φιλεῖ), writes Heraclitus of Ephesus.* But a philosopher must try to reach it somehow, because truth is very sublime.

Plato used the "myth of the cave," in which he describes a scene of slaves chained in a cave, who are forced to watch a strange "film" of speaking shadows on a wall. They believe what they see is real until one slave escapes and discovers an unexpected world: what the prisoners think are people are only the shadows of statues of humans and animals being carried on the shoulders of real men and women passing by; the slaves were hearing only their voices.[5] The freed slave met the other side of things. Centuries later, the neo-Platonist Giordano Bruno of the Renaissance wrote *De Umbris Idearum* (The Shadows of Ideas), and indeed Platonic thought has also been revalued by some quantum physicists. The physical bodies that we can touch, see, and hear are only the shadows in the cave. Their fields, though they elude our senses, are in fact the true reality of the bodies. A researcher has to leave the cave in order to explore the other side of things.

Every physical body can be seen as an event that is constantly changing on the world stage, and the director of the changes is precisely the field, which the ancient sages identified with fire, a great natural alchemist. The quantum field is everywhere. The particles are not corpuscular, but local condensations of the field. Solid? No. They are quanta, but they are packets of energy of the field's vibrations. The protons

---

*According to Heraclitus, a Greek philosopher before Socrates, of the sixth century B.C.E., reality is a continuous becoming and everything changes all the time, except for the law of flux itself, that of the tension of opposites, for changing is always defined by two opposite poles. For Heraclitus, the symbol of this constant mobility is fire, which is the origin of the world, the *archè*. The fool believes the outward appearance of nature; the wise person knows how to penetrate through the law that governs the world to see that becoming is only appearance; in reality, "it is wise to say that all things are one."

are vibrations in the field of the protons, electrons in that of the electrons, and so on. It is revolutionary in the history of human thinking to imagine that the world is not built with solid bricks, but rather with vibration, energy. Matter is a particular vibration of its own field, which overturns everything so far studied in school.

Since our childhood we have wanted to humanize the world, and we imagine even the microscopic driving energies of life as solid objects. But things are not like that. The Italian doctor and physicist Massimo Corbucci writes that the atom is an abyss filled with electrons and the particles of the nucleus.[6] The harder you search the abyss, the more you realize that mass itself does not exist. What exists is a game of attraction and repulsion (therefore a balance) between different polarities of charge, between "breathing emptiness."

The field is pulsation in the emptiness, that is, vibrating emptiness, a pulsating vacuum. The particles that make up mass might actually be disturbances of the field, *ripples in the vacuum*. We are not far from the discourse of the strings. Now consider that the first description of matter, as being like "the crest of a wave, curling like the sea," was written as early as the hermetic treatises of the second century C.E.! It is only these disturbances that are perceived by the senses, which then turn them into perceptions—visual, tactile, auditory—namely feelings from forms, bodies, heat, sound, light.

What appear to us as particles are probably field fluctuations, in which some of a field's regions oppose one another (for example, the protons and the electrons). In physics' "double slit" experiment, an electron sent toward a plate with two parallel slits close to each other passes through both simultaneously, suggesting that the electron is traveling more like a wave than a particle. Actually, an electron can be in either wave or particle form, a variation of field fluctuation.

During our journey, we will discover further that the fields of physical bodies have extraordinary properties, that they are "organized masses" and that to date nobody has been able to uncover what organizes them and how. The physical, chemical, and biological sciences continue to largely ignore these questions. In fact, the field may not only be the result of what happens to mass, but rather the director of what happens to mass. To begin to understand how this can be, we are aided by the concept of *morphogenetic fields,* which offer us insight into fields with organizing disposition.

The Maps of Things

The existence of morphogenetic fields was postulated by a group of botanical embryologists in the past century in order to explain the growth processes of plants and animals, and the differentiation of their individual parts. According to their concept, the morphogenetic field may have informational characteristics that contribute to invisible planning, which gives form to the organisms as they develop. They may also help explain the ordering functions responsible for group actions and behaviors in many animal species. The raw building material remains the same; what changes is the design itself: it is this that "decides" shape, proportions, and limits with respect to growth. Only the morphogenetic field can explain why a person's arms and legs are different, despite the fact that they contain the same proteins encoded in the same genes.

One of the first to describe ordering fields was Harold Saxton Burr, who taught anatomy and neuroanatomy at the Yale School of Medicine. For at least two decades, Burr conducted research into the shapes of plants and animals, and also on hypothetical living fields that he called *vital fields* (V-fields). Each organism follows a pattern of planned growth, led by its electromagnetic field. Burr discovered, for example, that the electric field of a sprout has the shape of the adult plant. In an unfertilized egg, he discovered an electrical axis corresponding to the future orientation of the adult brain, serving as a guide to place the cell in the right place.[7] According to Richard Gerber, "It is highly likely that the spatial organization of cells is intended to be a three-dimensional map of the finished version: this map or matrix is a function of the energy field that accompanies the physical body."[8]

Burr was convinced that the fields could dominate and control the growth and development of every living form. He writes, "The molecules and cells of the human body are constantly being demolished and rebuilt with fresh substances from the food we eat. But thanks to the controlling V-field, new molecules and cells are rebuilt as before and are arranged in the same way as the old ones. When we meet a friend whom we haven't seen for six months, there is not one molecule in his face that is the same as it was at the last meeting. But, thanks to the controlling field, the new molecules are placed exactly in the old familiar layout and so we can recognize his face."[9]

Biologists are struggling to explain how our bodies maintain their shape despite the continuous replacement of substances. The particle affects the field, but it in turn is conditioned, points out Burr. "The design and the organization of each biological system are determined by a complex electrodynamic field which dictates the behavior and the ordering of components. It has correlations with growth and development, degeneration and regeneration and orientation of the component parts of the entire system. It can control the movement and the position of all particles within the entire system . . . Science believes that the electrical variations in living systems are the consequence of their biological activity, but I believe that there is a primary electrical field in the living system that is responsible."[10]

When Burr talks about forces, he imagines "superregulatory systems" governing physiology. According to him the condition of the mind influences the state of the field. These words sound like Buddha's: we become what we think. For Burr life does not happen by chance, but is rather the result of an organization delivered through electrodynamic fields that rule the positions and movements of all particles: "Vital fields impose a plan and organization of the material components, throughout the constant changing of all the living forms, forcing an acorn to grow until it becomes an oak, and only an oak. . . . Vital fields are influenced by larger fields in which our world is included (solar spots, for example), subject to a higher authority that forces them to change in various ways."[11]

The experiments conducted by our research group (see chapters 5 and 6) also suggest the existence of informed structures, which are able to build and organize physical bodies and put them in communication. But these structures are invisible, not perceivable with the naked eye or with equipment. And there we run into the limitation of current science, which is *almost* a certainty of knowledge. *Almost* because the senses are subjective and fail to capture dimensions different from our own: parallel universes are perhaps only one step away from us, but they may as well not exist. What exists *for us* is all that exists, at least as far as the logic of senses. Reality for us is all that we imagine.

Imagination draws the limits of our world. Ancients depicted the earth as flat, as was suggested by the senses. Today we can think of the earth in its roundness because we have seen the curvature of Earth from space. However, it is with difficulty that we imagine the solar system, especially the farthest planets. The galaxy is unimaginable, even more so, the universe. Distant galaxies are billions of light years away from our understanding. How can we imagine billions and billions of miles? Consider how the ancients thought of a fixed Earth at the center of rotating spheres. It took Galileo's telescope, the Copernican revolution, and satellites to replace this picture of reality. And we still don't know if our new images are the right ones . . . but this is another matter.

Under the microscopic lens, we have the same dilemma. Where does the world end? In quarks?* Beyond? The limit has been moved so many times! Research into the components of matter has involved generations of physicists who always review the previous theories. At the beginning of the nineteenth century, experiments carried out by Dalton† suggested that everything was made up of atoms and nothing else. But before the century ended, Thomson‡ discovered the electron; from there on, during the early twentieth century, physicists described all the components

---

*Quarks are the elementary components of the protons, neutrons, and all the other particles sensitive to strong interactions. We presently know six distinct types, which rotate on their own axis in one direction or the other (spin) like any other particle.
†John Dalton (1766–1844) was an English chemist, meteorologist, and physicist.
‡Sir Joseph John "J. J." Thomson (1856–1940) was a British physicist and Nobel laureate, credited with the discovery of the electron and isotopes, and the invention of the mass spectrometer.

of the atom. The particles seemed to be the new frontier, but then came Paul Dirac, who proposed the idea of antimatter. He was mocked for thirty years until antiparticles were discovered, and the scientific community tried to correct its mistake and the insult it caused by awarding Dirac the Nobel Prize. Fortunately for him, Dirac was still alive. Then quarks were discovered, and once again the frontier was moved forward. Physicists constantly change the image of the universe, and sometimes they discard it completely to start all over again.

The progression of numbers is an example of how the world has only limited representation on the mental screen. If I read 0.1, it is easy to imagine a tenth part of something, one of ten slices of a cake. But with 0.0000001 the mental effort is enormous. Imagine if there are tens of zeros after the decimal point! The most famous irrational number is π, academically approximated to 3.14: an understandable number that in reality would be 3.14159265358979323846. I wonder how many people even read all the numbers one by one; this confirms how useless it is to try to imagine something beyond our limits.

We are dealing with a world of representations suggested by the senses and the imagination, not a sound foundation on which to base dogmas and doctrines. Nothing is certain. Objective reality is unattainable. What shall we do? Stop searching and abandon this powerful passion? No. We should extend the research field to regions forbidden to the senses, into the void, and redefine what our senses declare to be "empty."

2 ⚬━• THE LIVING VACUUM

Solids and Voids

Let's start with a simple consideration, even if it is trivial: the world is interchanging solids and voids. Solids are the bodies, voids the spaces, even in microscopic cells, molecules, atoms, particles, and so on. Between the electron and its nucleus is an immense vastness of space: matter is almost empty. Do you know to what extent?

Imagine enlarging the world a million times, with bacteria measuring more than a yard and a section of a hair being a hundred yards in diameter. Even at that magnification you wouldn't be able to see a hydrogen atom, which is the smallest of all atoms. If everything is enlarged even more, sufficient to make a tennis ball as big as the earth, the world would be a hundred million times bigger. At that point the atom would be visible. If we then enlarge the atom until the proton becomes visible, its electron (which would still be invisible) would rotate at a distance of a hundred yards. Such is the relationship between solids and void in the field of combined matter; mass is almost completely empty.

Pragmatically, our senses decide what is void or solid: what they are able to perceive is solid, and everything else is considered empty. So *void is in fact the absence of perception, a vacuum of phenomenon*. The problem of the "in-between" or "what is between things" is one of the oldest questions among philosophers, physicists, and scientists. Void is matter.

If you don't understand it and you only trust the senses, you can't go further. As long as science doesn't investigate other aspects of matter, it will only get partial results. Some scientists consider the world to be a giant hologram,[1] while others talk about the universal network in which the nodes are the bodies (the solids) and the threads the relationships between them (the voids).[2] There are endless interactions between elementary entities, whose individuality is dissolved in the totality of the process. Life develops through information networks capable of self-regulation and self-organization.

For Ervin Laszlo,* one of the world's leading system theorists, life is a network of connected relationships: "There are significant signs of the existence of a subtle yet efficient field that connects everything and every event." He adds, "In addition to the electromagnetic field, the gravitational field, and the various quantum and nuclear fields, there is also a field that relates and connects all things that exist and evolve in space and time."[3] Only a united field can explain the connections between bodies, which—far from living isolated from the environment—are involved in the production and the transformation of the immense network that is always rebuilding itself. This concept echoes that of Italian philosopher, mathematician, and astronomer Giordano Bruno (1548–1600), who spoke of the "countless invisible threads connecting everything in the universe."

The same can also be said for every individuated being: there must be a single field that organizes its shape and structure, regulates its metabolism, and enables it to communicate with other bodies and with the environment. This field cannot be solid, for if it were we would have already discovered it. It is what seems "empty." However, though it is void of phenomena, it is not void of essence.

We are accustomed to the vacuum as a background for the solids that we perceive. What would the world look like if we reversed the perspective, if we could focus on the vacuum background as solid? We can

---

*Ervin Laszlo, twice nominated for the Nobel Peace Prize, is also founder and president of the prestigious Club of Budapest, an international think-tank organization integrating science and spirituality.

be aided by a look at what is known as the "Rubin vase" (fig. 2.1), an example of the paradox of reality. Within the drawing are two images that emerge from a background that can alternately be black or white. Against the black background a white vase is visible, while against the white background two faces in profile are seen, almost kissing. Which is reality: the faces or the vase? Maybe both, but never together, because the brain needs a frame of reference in order to know how to grasp things in relation to others. Everything is relative, and our perceptions arise from the differences and contrasts; in this world all that is homogenous is invisible.

The world is based on dualism: if there is white then there is black, if there is night, then there is day, and so on; for everything there is an opposite, and becoming arises from the alternation of these possibilities. But from the pair of opposites, one is chosen each time to be, while the other is suspended and dies. Only in this way can the plot evolve. This is why Cain kills Abel, Romulus stabs Remus, and Castor dies but Pollux lives. Separation and exclusion allow evolution.

The vacuum is responsible for life: a world without vacuum would be an incomprehensible mass of dense matter. It gives identity to bodies;

Figure 2.1. The "Rubin vase"—faces or a vase? Both maybe, but never together, because the brain needs references and knows how to grasp things only in relation to others.

within matter it creates gaps and spaces that carve out atoms and molecules. It sustains distances in such a way that there is always balance in form. Like in Genesis, creation happens from separation. The vacuum is responsible for widening and separating matter. Matter combines in mass, and also fragments and dilates to create space.

The Latin poet Tito Lucrezio Caro writes in the first book of his *De rerum natura* that "the world is not all formed and forced by compact matter: there is a vacuum in things." He adds, "Vacuum from which originates the movement of all things."[4] He says that in the beginning there was matter, which was deprived of vacuum, so it was motionless and eternal. Then, becoming fragmented, it gained forms and the possibility of movement. If matter had remained compact, nothing could be distinguished and it would be, as Lucrezio writes, a "dark desert of blind atoms."

Vacuum *Non Datur*

A handful of seconds . . . time to grab the manuscripts . . . knocking over chairs and dashing down the stairs with a sudden roar before disappearing into the dark street. In haste, the Bolognese landlord threw on a cloak to give chase, but the monks from Rome had already disappeared without paying the bill. It was awful when Tommaso Campanella (1568–1639) discovered that the books he was planning to publish in Padua had been stolen. Discouraged, he realized that there was only one thing he could do, and he would do it: write them again. In 1595 the court of the Roman Inquisition incriminated him using proof from his original manuscripts in Latin, stolen that night by fake monks. Among those originals, the *De sensu rerum* was never found again, but there is an Italian version that Campanella rewrote during his twenty-seven-year stay in the prisons of Naples, published in 1604.[5]

Almost in anticipation of quantum physics, brother Thomas writes that vacuum is a divine creature, a continuity that holds the universe together. In the void there is *something* that gives life to the bodies before their creation, so much so that some Arabian philosophers thought emptiness was God.* These were ideas that few knew of and that certain ecclesiastical authorities tried to suppress by persecuting any who tried to understand.

In those years, ten years before the pyre, Giordano Bruno published the *De rerum principiis et elementis et causis,* in which he writes that emptiness is filled with a *spiritus* or *virtus* that occupies the space of matter: "Spirit is a mobile substance of which is transmitted to the bodies of every type of local movement, that is the father of every impulse that Plato instead calls soul. . . . Bodies don't have the capability of moving but they receive it from the spirit in them . . . we have to conclude therefore that in the spirits every strength and virtue is implicit."[6]

Virtus does not just refer to virtue but also to strength, power, and energy. In a metaphorical sense, it is wonders and miracles, as described in the Gospels: the energy that works miracles. Let's reread the episode of the healing of the woman suffering from hemorrhage, when "immediately the source of her blood dried up, and she felt that she was cured of her disease. And Jesus, who knew that virtue had gone out of him, asked: 'Who touched my robe?'" (Mark 5:29–30).

If we understand the true meaning of the well-known proverb *in medio stat virtus* (virtue is in the middle), we know that not only does it mean that moral virtue lies in the center or the middle way, but also that in the empty spaces of matter there is a miraculous virtue. We pay

---

*In the ninth chapter of the first book he writes, "All the entities abhor the emptiness that there is between them and so, with a natural impulse, they run to fill it to maintain their integrity and to enjoy a mutual contact. . . . So it is necessary to state that the world is a sentient animal and that it enjoys all the parts of the common life . . . and this happens when between the bodies partially intercepted emptiness remains." In the twelfth chapter, "And I am certainly of the opinion that space . . . attracts bodies to itself, not with instruments, but with appetizing sense, because it still has the power of being and sense of being and love of being such as that God has done it; a few Arabs believed that space was God because He supports everything and He is against nothing and everything He receives benevolently. . . . I certainly admire its nobility, but that it is God I don't believe. But I know well that it [the void] is the basis of every creation and that it precedes every created being, if not in time, at least in nature and origin, because if the world was created, Averroes says that it is necessary to confirm that before that there was emptiness. . . . Or if the space it is indeed the divine creature . . ."

attention to the deeper meanings of words of wisdom, maxims, and proverbs. Even Leonardo da Vinci alluded to the existence of virtue *between* things as a force when he wrote, "The being of nothing is the greatest and what is *in between* things of the world is supreme."[7]

In vacuum there is a principle, which connects the bodies that seem to emanate from it. Its nature is unknown, but it is known that this untouchable, sacred, and indispensable thing achieves miracles and transformations. Giordano Bruno says "that compound formed by the spirit and the few subtle substances that emanate from the bodies pervades the surrounding space and that is the principle on which physical and magical operations are based."[8] The Renaissance neo-Platonists understood that to work miracles one has to act on the field of the substance, the part that seems empty and contains information, called virtus. The initiates knew this concept, as did the founding scientists of mechanistic thought such as Newton, who concluded in his *Principia mathematica:*

> Here it would be appropriate to say something about the Spirit, or ether: most subtle, pervading all solid bodies and hidden in their substance; by the strength and action of this Spirit individual particles of different bodies mutually attract one another at minimal distances, and cohere if contiguous. . . . In animal bodies it activates all sensations and moves the limbs, vibrating and spreading from the external sense organs toward the brain through the nerves and then from the brain to the muscles. But all this can't be expressed in a few words nor are we supplied with sufficient experiments to accurately determine and demonstrate the laws under which this electric and elastic Spirit acts.

The first page of the same book explains, "Here I don't deal with that medium—given that it exists—that pervades freely the interstices between the parts of bodies."[9] He implies, without saying so directly, after Giordano Bruno, Campanella, and others have been condemned.

It occurs to me that the "medium . . . that pervades freely the interstices between the parts of bodies" and the "most subtle [Spirit], pervading all solid bodies and hidden in their substance; by the strength and action of [which] individual particles of different bodies mutually attract one another at minimal distances, and cohere if contiguous" is the field. The vacuum is not empty; *vacuum non datur,* writes Girolamo Cardano (1501–1576) in Pavia, "emptiness doesn't exist."* It is not empty, for it contains "the principle on which physical and magical operations are based," which is nothing other than the information inscribed in the field. What kind of information?

In Medio Stat Virtus

The secret of the "virtuality" of masses has been known since the ancient times. As we have seen, among the first to speak of it was Parmenides of Elea in the sixth century B.C.E. He referred to the unbroken complexity in the emptiness between things as *being* and the solid bodies as *nonbeing.* Thus, we can think of that which is solid (mass) as nonbeing and the void (the area of a particular phenomenon that we are about to discover) as being. Parmenides favored the world of the void because he knew that is where things originate and from where they are governed. Centuries after Parmenides, Campanella spoke of the void "that precedes every created being, if not in time, at least in nature and origin."[10] Observe this point again in the following excerpt from the twelfth fragment of Parmenides:

> The narrower bands were filled with unmixed fire, and those next to them with night, and in the middle of these rushes a portion of fire. In the midst of these is the divinity (daimon) that directs the course of all things; for she rules over the painful labor and propagation of all, driving the female to join with the male, and the male with the female.[11]

---

*Girolamo Cardano was a great scientist, medic, alchemist, mathematician, physician, and engineer of the Italian Renaissance who wrote about medicine, philosophy, mathematics, natural science, law, astronomy, morality, divine arts, and alchemy. He was also incarcerated for heresy.

This leads to the following reading: in microscopic emptiness (between atoms, particles, and so on)—that is, *in between* matter—there is a vital energy like fire (the fire of Heraclitus). This is the most ancient intuition of a field with information in the spaces between things, information (the virtus of Giordano Bruno) that controls matter constantly. It is a principle of "labor and propagation" because it is from the field that things originate.

Being that still has to be determined is the "unique and infinite space" that Giordano Bruno wrote about, or the "pure matter" of Severi and Pannaria. Modern physics considers vacuums the source of all fields and all forces, including gravity and electromagnetism and the strong or weak nuclear interactions. Emptiness is the matrix.

Parmenides preached a move away from knowledge based on the senses because he considered it false. He thought that the world of the empty is the only reality, the being, while the full, the nonbeing, is illusionary, real only for the senses. All the major thinkers have rightly perceived that bodies are illusions, representations of something else, and what we call existence is like a dream. But this secret has been hidden, transmitted only to those able to accept the terrible truth that we live in a fiction.

Buddhists call *sunyata* (emptiness), the ultimate reality of things, "living emptiness" that gives rise to all forms. According to a Buddhist sutra, "Form is void, and void is indeed form. Void is not different from form; form is not different from void. What is form, that is void, what is void, that is form."[12] Like the Hindu *Brahman* and the Chinese *Tao*, sunyata is emptiness with infinite creative potential, similar to the quantum field of subatomic physics, in which the particles "are only concentrations of energy that come and go, thereby losing their individual character and dissolving in the subjected field."[13] "The field is the only reality" said Einstein, in a spirit similar to much of Eastern mysticism.[14] And, in the manner of the ancient philosophers, Giuliano Preparata* wrote "void is everything."[15]

---

*Giuliano Preparata, professor of theoretical subnuclear interactions at the University of Milan, was one of the world's leading exponents on QED (quantum electrodynamics), along with the founding fathers of quantum mechanics, Werner Heisenberg, Paul Dirac, Ernst Jordan, Wolfgang Pauli, Enrico Fermi, and Richard Feynman.

Two thousand years ago Lucretius stated that all things originate from emptiness. Why did physics become interested in emptiness again after centuries? It became interested because any quantum event originates from there and returns back there. "Today some physicists," writes Preparata, "really have begun to think that the fundamental particles spring from the field."[16] Let's remember this point because it is one of the fundamental theses of this book. The quantum emptiness is not so different from pure matter, the Great Mother.

Reality is transient and illusory. Vacuum is not empty. The field is the theater of exchanges. There are "virtues" in the field—information—"the principle on which physical and magical operations are based." The information of the field is essential to life and hides the secrets of existence. How can we access it?

A Key in the Puzzle

Even if we accept that all the matter that we *don't* perceive does exist and is present, we still need a way to get access from the world of the senses to the elements that can help tell us what lies behind things. It is like a crossword puzzle, with the white spaces to be filled in and the black spaces that cannot be filled. The black spaces are the other side of things, while the white are the little that we perceive and understand; and as knowledge grows it will fill the white spaces, as is happening while I write.

Attention needs to be paid to the black spaces though, just as in a crossword, it does not matter so much what is inside them; rather the key is to understand why they have their position, why they are there at all. The position of things is important. Try to leave a suitcase in the middle of the hall of any train station. If it is leaning against a wall, few will notice it. If it is left in the middle, however, it will sooner or later be viewed with suspicion. If it were placed near the front row of the ticket counters, where many people are standing in line, no one would notice it until closing time. More than the color or the content, the location is what makes the difference; it determines different plots in the film in which it plays the main role. We are of the same matter as that suitcase, and we carry certain rules within us, which means that we have at hand the evidence we need to understand the laws that govern the unknown behind things.

Research that carefully analyzes matter in its elementary components is confined to filling the white spaces, to understanding how the boxes are made. It can only become complete by extending itself to the black spaces, by constructing drawings in which the black spaces constitute the integrity of the composition. We need to pay more attention to synthesis as well as analysis, to realize that a "background noise" is not always without meaning, and that a particular, seemingly useless, detail could be a valuable piece of a larger mosaic if we are able to observe its integral significance. What matters is what holds the particles and molecules together, what organizes them into bodies. The different positions of the black may seem random, but knowing how to look, we discover alternative rhythms, languages, and codes. They have a meaning. That is what matters: not how the boxes are made, but the order of the boxes in between them; there is a code hidden in the alternatives.

If it is written in the DNA of your daughter that she will have blue eyes, her genes will print the very pigments of the iris. But we have yet to fully understand the fact that your son's long arms can be in the exact same proportion to his legs and chest as his grandfather's. Who and what tells the cells to grow only to a certain point and then stop? How is it communicated? Who puts the molecules of a substance in a certain way, and only in that way, so that it becomes quartz crystal instead of a shell or a fossil? Where is the pattern that allows the silica to become crystal; who directs construction, who teaches geometry? These are the problems for which we are seeking answers.

First, though, we require a pause, brief but intense, to explore the world of water and its mysteries, for the unusual properties attributed to water offer us some important clues.

3 ⊷ AQUATIC INTERLUDE

The Strange World of Water

The unleashed power of the great wave in Hokusai's (1760–1849) *Mount Fuji Seen Below a Wave at Kanagawa* is depicted as foamy paws ready to hit tiny mariners, serving as a reminder that we live in a world almost completely aquatic. The snow-capped mountain is nothing compared with the disturbing sea. Three-to-one is the proportion of water to earth on the planet: three-quarters water and one-quarter earth. The human body is also three-quarters water: our body and our planet are governed by water. Water is everywhere, not only in the sea, in the rivers, in the lakes, or in the clouds, but also in the air as vapor, in our breath, and on the surface of any physical surface, even if it may appear dry.

According to Thales, one of the first Greek physicists and philosophers, water is ἀρχή, the basic cause of things; everything is born and derived from it. Mysterious, humble, and powerful, water has played primary roles in the history of humanity. In 1781, British scientist Henry Cavendish understood that water is made of hydrogen and oxygen. However, it has only been known since the beginning of the last century that water is composed of two hydrogen atoms and one oxygen atom. The two hydrogen atoms are 0.95 Å (angström)* away from the

---

*The angström (Å) is a unit of length equal to one hundred-millionth of a centimeter, 10^{-10} meter (or one ten-thousandth part of a micron), used mainly to express wavelengths and interatomic distances.

oxygen atom, and together form an angle close to 105°, thus the molecules of water take the shape of a letter V (fig. 3.1).

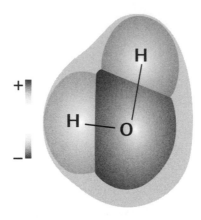

Figure 3.1. Structure of a water molecule; magnet dipole

The structure of the water molecule can be likened to a magnet, with the two hydrogen atoms positively charged and the oxygen negatively charged. This allows water molecules to interact with each other, establishing an attraction known as a *hydrogen bond* between the hydrogen and oxygen atoms of different molecules. In one drop of water, there are billions of these bonds. Such relationships between molecules are called "interactions at a short range" and are responsible for the network-like structure of water.

Brian Josephson, winner of the Nobel Prize for Physics in 1973 and professor at Cambridge University's Cavendish Laboratory, writes:

> Scientists have little knowledge on the topic of water and they tend to have a naive vision: liquid composed of H_2O molecules more or less isolated in movement. In reality, water as a phenomenon is a lot more complex, with single molecules grouped temporarily to form a lattice structure. It is not at all surprising that these molecules can interact, thus giving rise to a mechanism that allows water to have memory, although the existence of such a mechanism only rings true to well-informed scientists who don't underestimate the possibility of its existence.[1]

The most interesting property of water is that it is able to retain information it receives. Let's try to understand how this is possible. If the network phenomenon is true for water and ice, it is not so for water vapor, where every molecule moves on its own: evaporation dissolves the hydrogen bonds between molecules. While on the one hand water molecules are attracted to one another because of their charged status, on the other hand they move away on their own; they cannot stand still. A stretch of still water may appear to be motionless, but this is not so. Its surface is in constant turmoil due to microscopic movements, known as *Brownian motion,* which increase with temperature. Water moves because its molecules oscillate at certain frequencies.

If we imagine ourselves as a tiny atom and we dive into a glass of water, what would we see? Among the many molecules (huge now to our eyes), gigantic icebergs made of dense water would appear. They are called clusters, very big masses of water molecules. Thus, water can exist in two phases, one at a higher density and one at a lower one: "more solid" water and "more liquid" water.[2] The clusters are linked together through "knots or nodes" (hydromagnetic junctions of a fluid crystal type), which make them similar to crystals, and indeed water can behave like a liquid crystal.* This helps to explain some unusual properties of water.

Water is an important solvent because the magnetic structure of the atoms dissolves the negative and positive polarity of the solute, for example a salt, which turns into *ions,* molecules missing some electrons. The ions are "solvated," that is, the water molecules surround them completely, like a blanket over an infant. The two types of molecules, solute and solvent, interact and reorganize the structure of water. The water is the same from a chemical point of view—H_2O—but from a chemical-physical perspective it has different properties: its molecules are arranged in a different way than before, due to received information that is retained. When you dissolve something in water, water molecules group together (with networks of hydrogen bonds), guided in a way that

---

*A *liquid crystal* is an intermediate state of matter between solid and liquid, with several properties of both crystals and liquids that, for example, allow a variation of the molecular orientation with the application of a magnetic field.

is imposed by the solute. When you pour sugar into a cup of tea, water molecules arrange themselves in a way to accommodate the sugar, and this arrangement is specific for sugar. Another arrangement will be made for salt (provided you like salty tea), and so on. In other words, what is dissolved is what determines the reorganized structure of the water.* Now, let's tackle a question that has been debated for many years: is there only water in homeopathic remedies or something more? Do they have a physical basis, indicating that we should consider them as drugs?

The Enigma of Water

At the end of the eighteenth century the first homeopathic remedies started to appear, made by Samuel Hahnemann. He didn't have tools or machinery of any kind; he had only dilution substances, water, and alcohol to mix in glass bottles, as was customary among German doctors in the country at that time. Like an alchemist of the past, he prepared tinctures, with plant materials, animal organisms, or minerals dissolved in alcohol, then diluted them with water in successive steps, each time shaking the mixture a hundred times. Such shaking is termed *succussion or dynamization.* Without it the preparation is only a diluted substance, not a homeopathic remedy. According to Hahnemann, homeopathic preparations must be dynamized against "a rigid but elastic surface, such as the leather cover of a book."[3]

In homeopathy dilutions are most often used in a ratio of 1:100 or 1:10, then in centesimal or in decimal sequences. Starting from a liquid tincture diluted to 1 percent in weight or in volume, after succussion it becomes the first centesimal Hahnemannian (1 CH), adding 99 percent of water and again succussion, then it becomes the second centesimal (2 CH), and the process continues until the desired dilution is obtained. At the third centesimal, the solute concentration is a few parts per million,

---

*This model is confirmed by X-ray diffractometer data (A. H. Narten, "Levy," in: *Water a Comprehensive Treatise,* F. Franks ed., New York: Plenum Press, 1972) and by Raman spectroscopy (G. E. Wahrafen, "Raman Spectral Studies of the Effects of Urea and Sucrose on Water Structure," *J. Chem. Phys.* 44 [1996]: 3726).

meaning that the solution is almost entirely water and the pharmacological effect of those few molecules of a substance is zero. After the twelfth centesimal the absence of solute is total, and there is no trace of the dissolved substance. The few molecules present in homeopathic preparations after the 3 CH are not sufficient to act pharmacologically: the introduced substance remains only a "memory." Even though from a chemical point of view the remedies are only made of water, the therapeutic effects of the homeopathic remedies seem to relate to specific configurations of water molecules. To better understand this we need to shift our thinking and reasoning from a chemistry bias to a physics bias; this is the stumbling block for those who negate the efficacy of homeopathic dilutions.

There are two different angles from which to tackle the problematic issue of homeopathy: clinical and physical. I don't mean clinical efficacy, which may depend on factors not fully controllable (starting with the experience of the physician). What is relevant here is the physical aspect, understanding whether after homeopathic dilution water is truly different, having gained therapeutically powerful information. Let's examine the theories proposed so far and the chemical and physical data that support them.

Chemical Molecular Hypotheses

The Greek physicist George Anagnostatos formulated the first model in 1988 (clusters theory). In it, water molecules wrap around the dissolved substance like a cast. When the solute is progressively eliminated (through further dilutions), the water molecules eventually collapse, filling the space left empty by the solute molecule. This leaves a mold made of water, which is the copy of the lost molecule of the entirely diluted substance. This is like the process of casting a bronze statue, in which first a model is made of another material such as wax (the molecule dissolved in water), then it is wrapped in plaster, which becomes the "negative" of the statue (the mold of water molecules). The wax is then removed so that molten bronze can be poured into the mold to obtain the bronze statue (the mold collapses, becoming a copy of the substance that had been dissolved and taken away).

Water with such a changed structure attracts other molecules to form new casts, which, upon further dilution, become empty niches and then collapse again immediately. The phenomenon continues to multiply until it involves other molecules. Once the original solute molecule has been copied, the mechanism is started, and the water continues to multiply copies. According to Anagnostatos, the physical structure of water is gradually changed until it ultimately completely takes on the spatial configuration of the diluted substance.[4] The altered configuration remains even when the solute is completely eliminated, and indeed it is this absence that enables the process to take place.

In 1996, Israeli doctor Dorit Arad observed complex molecules dissolved in water via radioisotope marking.* He discovered that the water casts around dissolved molecules were selective and specific to a substance's most active site (which is the portion of the molecule that has the greatest pharmacological effects). This means that the water molecules are perhaps attracted by differences in potentiality in the active site. The water builds its models only on this part of the molecule, not on those that are of minor pharmacological importance.[5] This seems to suggest that for a complex molecule, the water is able to select the most useful parts, from an informational point of view. The memorization within the water would then selectively happen under the guidance of a regulating system intrinsic to the water itself.

Chinese researcher S. Y. Lo proposed a different model. In his model, in very high dilutions the ions present in water generate an electrical field by their movement; this in turn produces aggregates that are stable in water. For this to happen they must occur at speeds much faster than that which destroys them.[6] It is possible to increase the number and the dimensions of the aggregates by simply shaking the solution (succussion). Proof of their existence was derived by measuring the transmittance of ultraviolet light.[7] S. Y. Lo had noticed in 1996 that in homeopathic dilutions the absorbance of ultraviolet light is reduced.[8] He thinks that the movements of the aggregates of water molecules, which are tiny magnets, are responsible for the so-called memory of water.

Isotopes are similar elements but with a different atomic weight; radioisotopes are isotopes made radioactive.

Further confirmations came from the electronic microscope, which showed aggregates of water molecules in the form of small sticks.

Unexpected Coup in the Nightclub: The Physical Hypothesis

What we have discussed so far is the chemical and molecular hypothesis. The proposal of theoretical physicists Emilio Del Giudice and Giuliano Preparata at the University of Milan is instead of a quantum type. In quantum physics particles as well as all bodies constantly fluctuate, because nothingness doesn't exist and everything is in continuous interchange. From a physical point of view, a substance that vibrates at the same frequency as another becomes a copy of it, even if it remains chemically different.

In water each molecule typically vibrates on its own, independently and in seemingly disordered fashion, just like when people are dancing in a nightclub, one here and another one there, each in his or her own way. But if a substance is dissolved in the water, it can instigate a coup, as if the music suddenly stops in the nightclub and the speakers instead broadcast classical ballet music while the youngsters get together, arm in arm, dancing the same steps at the same pace, like a ballet troupe. It is as though a DJ from out of town has taken over from his colleague and imposed his own music on the dancers. Similarly, the diluted substance imposes its own vibration—its information—on the molecules of water. Thus the information of the solute is imprinted in the water because it modifies the oscillation of the water molecules. Unlike the chemical models described above, in which molecules are copied, in this model the molecular rhythms of the solute are impressed on the water.

Before, the water molecules were each vibrating on their own, but in the presence of the solute they begin to fluctuate in phase. *In phase* means that the water molecules now oscillate at the same rate, acting as if they were a single molecule. In homeopathic dilutions the molecules of the solute (the substance that is dissolved in water) are able to capture the vibrations of the water and make them oscillate at their own pace due to certain phenomena of electromagnetic interference. In order to explain those phenomena, we need to begin with understanding the concept of a *ground state* (GS). A quantum system is stable when it reaches—which it does sooner or later—its "state of minimum energy," its ground state. A system achieves the ground state through fluctuations that dissipate any surplus energy. The GS of a system is similar to what happens when a car reaches the speed that allows for the optimal balance between fuel consumption and stability.

The state of minimum energy is the quantum vacuum. Left alone, a quantum system tends toward the vacuum: this is how matter can interact with its field. If matter comes to interact with its own electromagnetic field, it can reach high levels of coherence, that is, it becomes "highly ordered."* The system's state of minimum energy becomes a *coherent ground state* (CGS) when molecules and fields oscillate in phase with each other. When matter and field oscillate in phase, they produce high levels of consistency and coherence. Areas where the molecules oscillate in a coherent way are called *domains of coherence*. These domains of coherence allow water to record information and "imitate" anything.

Bodies are defined both by short-range forces (molecular interactions) and long-range forces (field oscillations). Molecules that oscillate in phase can emit intense and penetrating waves. Oscillating in phase with their field, the particles vibrating at the same frequency generate waves of photons, which are rich in consistent particles.[9] When they oscillate in a coherent way, systems emit more intense signals, just like a laser.

Entering a coherent ground state (CGS), water can receive and emit signals. According to Preparata's "theory of quantum electrodynamical coherence in condensed matter," this applies to liquids and solids, in which atoms and molecules interact via both short-range forces (chemical bonds) and long-range forces (mediated by the electromagnetic fields).[10] What does this mean? The molecules generate a field, which in turn is like a conductor who, through long-range interactions, harmonizes the movements of all the molecules. The short-range chemical interactions are the dynamics of the system, but the long-range interactions retain control.[11]

This is the secret of the "memory of water": the coherence of its fundamental state, which enables it to convey information. All this can be

---

Coherence in physics expresses the level of order, which corresponds to a high degree of information.

explained by the theory of superradiance, formulated by Robert H. Dicke in the 1950s and revisited by the physicists Preparata and Del Giudice. Superradiance is a class of enhanced radiation effects in which disordered energy of various kinds is converted into coherent electromagnetic energy. Particles of coherent matter can produce amazing effects when oscillating with their electromagnetic fields. The secret is the oscillation. Oscillating particles generate a field that fluctuates in a constant way, typical of systems that are self-governing and self-maintaining. In other words, water is able to receive, retain, and return information because it fluctuates between coherent and noncoherent states. This enables water to be a medium for excellent communication.

The human body consists mostly of liquid, which is subject to the alternation of coherent states of water. If most of the molecules of this liquid lose their pace and their state of coherence, the consequence is a disorder in the information content being carried, which in turn can cause states of disease. Imagine the immense amount of fluid surrounding the cells (called mesenchyme) and within the cells themselves: like gigantic moving seas whose state changes continuously, just like water freezing, then becoming liquid, and then dispersing in the form of vapor. When dissipated, the bodily fluids collapse into a state of emptiness, losing coherence. Some system information is then deleted. Of course, the body's control systems work to ensure that this does not happen—like the antivirus systems of a computer. But if something escapes the control mechanism and manifests itself as a symptom, the introduction of water carrying the appropriate information can act on the body's fluids in order to help return them to a coherent state. This might be a homeopathic remedy or any other "informed" water.

Some Like It Hot

The first scientific proofs that homeopathic dilution is different from plain water have recently been obtained through nuclear magnetic resonance (NMR). NMR is a property demonstrated by magnetic nuclei in a magnetic field when an electromagnetic (EM) pulse is applied, causing the nuclei to absorb energy from the EM pulse and radiate this energy back out. The energy radiated back out is at a specific resonance frequency, which depends on the strength of the magnetic field and other factors. Recent analysis of nuclear magnetic resonance has confirmed that the water in homeopathic dilutions is organized differently than that in the controls.[12] In fact, the NMR techniques show that double-distilled water is quite another thing than the highly diluted homeopathic solutions, which show the organized presence of information.[13] Every time the chemical and physical parameters confirm reorganization, it means that water has acquired information via the solute or via the field.

The chemist and physicist Vittorio Elia of the Federico II University of Naples has discovered important variations in the thermodynamic profile of homeopathic dilutions using a technique known as microcalorimetry. Imagine a hollow sphere in which two tubes are placed: two different solutions pass through those little tubes and mix together. In their mixing, millions of molecules collide, generating a very weak heat. In microcalorimetry this heat is measured precisely. (The cells are thermostatically controlled so that the environmental temperature does not influence them.) If the heat measured is typical of water, it confirms that only molecules of water are present. If, however, the water contains different molecules (such as, for example, those of a medicine), the heat readings will be different, generally increased.

If homeopathic remedies were only constituted of water, their profile would be calorimetrically that of water (and those who consider homeopathic remedies nothing but water would be right). Instead, thousands of samples of homeopathic remedies analyzed by Elia over the years had heat signatures higher than that of pure water, demonstrating that there were other molecules present besides water.[14] This was true despite the fact that chemical analysis revealed only the presence of water. It is plausible that the excess heat is due to the water molecules organizing in a different way from before. This is what we call information. Thus, homeopathic remedies are not just fresh water; on the contrary, they are warmer. The homeopathic dilutions are also different in the sense of temporal evolution: the excess heat doesn't peter out in a short space of time, as it does for medicines, but instead increases with time.[15] In periodic measurements of the dilutions, the heat recorded is increasingly higher.

As difficult as it is to explain how water-based solutions can contain molecules that supposedly aren't there, this progressive calorimetric increase is even more surprising. Even undiluted water (nothing dissolved in it) but dynamized (through a process of violent succussions, as in the homeopathic procedure) increases in heat, which confirms that this mechanical action is necessary in the homeopathic preparation. This is not explainable via traditional chemistry, which says that this action itself is not sufficient to reorganize the molecular disposition of water. On the other hand, diluted solutions that are not succussed are completely different from homeopathic remedies. Succussion is therefore essential to modify the structure of water and plays a key role in the mechanisms by which information is stored.

It has also been shown that homeopathic remedies have a higher conductivity than undiluted water, that is, an increased ability to transmit electricity. Chemically, water is always H_2O, but physically it can differ. This is an undeniable phenomenon. There is now evidence from NMR, calorimetry, and measurements of chemical and physical conductivity showing that water that has received information is different from before.

More Breakthroughs: Electromagnetic Signals from Water

In 2009 the Nobel Prize winner Luc Montagnier* published a paper in which he officially demonstrated the existence of homeopathy, confirming that water, in homeopathic dilutions, not only keeps memory of the solute but also emits specific low-frequency electromagnetic signals that are both recordable and reproducible.[16] Let us see how.

Montagnier prepared homeopathic dilutions from water solutions of different bacterial DNA sequences. These samples were serially diluted

---

*Luc Montagnier, a French virologist, was a joint recipient of the 2008 Nobel Prize in Physiology or Medicine (with Françoise Barré-Sinoussi and Harald zur Hausen) for his discovery of the human immunodeficiency virus (HIV).

1:10 (0.1 + 0.9) in sterile water in tubes and then tightly stoppered and strongly agitated on a Vortex apparatus for 15 seconds. This step (succussion) was found to be critical for the generation of signals.

The tubes were read one by one using a device designed by Jacques Benveniste that which detected signals through an electromagnetic coil. This device was connected to a Sound Blaster Card that was subsequently connected to a laptop computer (preferentially powered by its 12-volt battery). Each emission was recorded twice for 6 seconds, amplified 500 times, and processed with different software to visualize the signals on the computer's screen. The main harmonics of the complex signals were analyzed by several types of Fourier transformation software.

In the dilutions from 10^{-7} (D7) to 10^{-12} (D12), he visualized positive signals in the range of 1 to 3 KHz. These signals signify new harmonics, but they were found only in the diluted and succussed solutions. Montagnier thinks that the emission of such waves likely represents a resonance phenomenon, depending on excitation by the ambient electromagnetic noise, and it is associated with the presence of polymeric nanostructures of definite size in the aqueous dilutions. The supernatant of uninfected eukaryotic cells used as controls did not exhibit this property.[17]

Montagnier thus confirms that homeopathic dilutions emit frequencies that are nonexistent in the initial undiluted solutions or in low dilutions (10^{-3}). Therefore, from the physical viewpoint, this water is different from what it was before. It was stated by a Nobel Laureate; Ipse dixit.

This emission of electromagnetic signals is not suppressed by enzyme (RNAseA, proteinase K), formamide, formaldehyde, or chloroform. The cations are able to reduce the intensity of the signals, while the range of the positive dilutions remained unchanged. However, heating at 70° C for 30 minutes irreversibily suppressed the activity, as well as did freezing for 1 hour at -20° or -60° C. The nanostructures emitting electromagnetic signals seem to have the same range of sizes as those originating from intact bacteria.[18]

The physical nature of the nanostructures that support the electromagnetic signal resonances remains to be determined, but it can certainly be ascribed to the above-mentioned clusters. The latter are indeed inactivated by sensible temperature variations, as observed by us, by Benveniste, and by others. Montagnier has discovered a novel property of DNA; that is, the capacity of some sequences to emit electromagnetic waves in resonance after excitation by the ambient electromagnetic background (and this can be related to the theory of biophotons emitted by the DNA observed by Fritz Albert Popp, which will be discussed later).

It seems possible to transfer the information between two different dilutions of DNA (one emit and another is silent: for example, 10^{-9} and 10^{-3}) when the tubes are placed side by side in a mu-metal box for 24 hours at room temperature. This crosstalk can be suppressed by interposing a sheet of mu-metal between the two tubes during the 24-hour contact period, pointing to a role of low frequency waves in the phenomenon.[19] The crosstalk is in accordance with what was observed by Schimmel, Endler, Popp, Elia, Citro, and others (see the foreword).

In 2010, Montagnier received a delegation from my Research Institute at the UNESCO Foundation in Paris.[20] In his office, located on the fifteenth floor and overlooking the Eiffel Tower, we exchanged information about our respective and very similar research studies. Starting in the fall, we began Montagnier's experiment and worked at it through the winter, until we were able to reproduce it.

Montagnier was a friend of the late Benveniste; moreover, his current assistant, Jamal Aissa, worked with Benveniste and had taken part in our first TFF lab experiment in 1992 in Paris (see chapter 6). Both researchers came from a medical background, but they also shared another, more compelling commonality: that they dared trespass the boundaries of chemistry to explore the physical side of things—the realm of electromagnetic waves, and a lot more. With his research, Montagnier closed the water memory circle that had been opened twenty years before by Benveniste and—because of the influence coherent states of water has on the health of the human body—paved the way to new and interesting future prospects for the early diagnosis of many important chronic diseases.

o—•

If we consider all the publications by serious researchers who are investigating the water phenomenon in the true spirit of research and not governed by other interests, the sterile and useless debate on homeopathy can come to an end, for they make it clear that homeopathic remedies are water with different physical characteristics, remedies with a distinctive preparation. They are based on the ability of water to reorganize according to the informational imprint of a solute. As we shall see in later chapters, the same principle applies even when we transmit information to water electronically.

Water behaves as a natural recorder and transmitter of frequencies. During our trip you will see that it is not alone and that all matter is capable of receiving and sending signals. This provides us with some important clues to the organization of the universe as we continue our journey into the other side of things.

4 •━ INTO NATURE'S NETS

Everything is alive, including objects. So instead of distinguishing between "things" and "living beings," we will speak of "cellular" and "noncellular" beings. During this journey we intend *things* to refer to bodies devoid of cells: particles, atoms, molecules, chemical elements and their compounds, also medicines, raw materials, stones, rocks, minerals, metals, objects, artifacts, houses, buildings, mountains, rivers, seas, regions, continents, planets, solar systems, stars, universes, and so on. *Organisms* such as plants or animals are, on the other hand, made of cells.

Creation is subject to planning. Something teaches the molecules of the crystals where to place themselves during their growth. This also applies to the human body and to any organism: cells follow an assembly scheme in order to know what place to occupy, in what direction to develop, and at what point to arrest their growth. They must complete their form. What inspires bees to build impeccable hexagons? This is not an easy task, as they are so tiny and are suspended in emptiness, without any scaffolding! Or do you think that everything happens by chance, a possibility that happens to repeat itself? Instinct perhaps? But what is instinct? The holder of the design or the design itself? Who directs complex natural processes? Who owns the designs?

Natural Schemes of Organization

In the history of scientific thought, knowledge has always followed two different paths: one is involved with studying the substance, the other the form. The first question was "What is it made of?" and the second "What is its scheme?" For a long time, substance prevailed over form. Then in the 1960s, the development of cybernetics, together with system theory, led to a resumption of the study of form and its patterns of organization, which in turn helped pave the way for computer science.

Nature inspires technology. A biological system is self-regulating and self-organized. "The pattern of life is a pattern of networks capable of self-organization,"[1] writes Fritjof Capra: the organizational patterns of living beings ensure that their components form an interconnected network, so that in observing life, we are observing networks of information in all directions.[2] Think about feedback loops, cyclical relationships in which networks are self-regulating: an error in the system tends to spread throughout the network, but the loops allow the information to feed back to the source to be checked so that the error can be corrected. Homeostasis* always respects the balance, like a marathon runner who, finishing the race, replenishes the fluids he lost along the way.

In the 1960s the laser was invented, an example of a structure that is self-organizing. Think about a light bulb, whose light waves are all chaotic, with different phases and frequencies. Now imagine being able to make them parallel and compact as a single beam directed from a single site. This is a laser, a self-organizational light that has become "coherent." It coordinates its own emissions, a principle that applies to all atoms or complex organisms. There are infinite "lasers" in the fabric of matter.

In the 1970s and the 1980s researchers from different nations

---

*Homeostasis is the self-controlling action of a living system, which continuously maintains its balance within its particular parameters of physiological stability, compensating for ongoing disturbances by external and internal factors.

studied the self-organization of living systems, and they saw the light, expressed by the theory of dissipative structures from Prigogine in Belgium, for example, or the Gaia hypothesis of Lovelock in England (the earth is a living system that is self-organized). Manfred Eigen, director of the Max Planck Institute in Göttingen, Germany, and winner of the Noble Prize in Chemistry, researched molecular self-organization, finding that at the apex of life there is molecular selection, by way of closed loops that are self-organizing. Chilean neurobiologist Humberto Maturana discovered that the nervous system always refers to itself: our perceptions do not represent any external reality, but are the results of interrelations in our neural network.[3] His collaboration with Francisco Varela resulted in the model of the autopoietic system, from *autopoiesis*, which means "self-creation."

In the autopoietic model, living systems are continuously governed by networks where each component participates in the transformation of others and the entire network produces itself.[4] This is not the invocation of transcendent forces, nor is it Cartesian mechanism. The focus of Maturana and Varela is concentrated on the relationship rather than the characteristics of the components: the black boxes, rather than the whites.

The organization of a being is the totality of its connections.[5] A network that is self-confined is an autonomous force field, where the elements can be adjusted to each other like reflections in the mirrors at the amusement park. It is a mechanism that is repeated in nature: the particle in relation to those that generate it, the network that regulates the components that in turn produce it, the field that maintains the particles that give rise to order, and the field that organizes the matter that maintains the field itself. In this type of relationship it is not possible to distinguish between cause and effect: which came first, the egg or the chicken? It is a relationship of two mechanisms, where each generates, sustains, and regulates another. And indeed, perhaps, it is not dualistic, but a single thing, as in the case of the wave/particle, something unique that we still don't know.

The Governing Code

A specific form becomes possible because of the relationships and the numerical patterns contained in the design. The same happens in the fractal geometry that was discovered by the French mathematician Benoit Mandelbrot. *Fractals* are geometric shapes, extremely uneven, each part of which has the same statistical character as the whole. The same structure (sometimes very complex) is repeated in every single part and in successive enlargements. In fractals, the form repeats at every order of dimension, like in Chinese boxes. A cauliflower provides a good example: there is a noticeable similarity between any piece broken off of it and the whole cauliflower; further subdivisions will produce minuscule cauliflowers. This self-similarity is common in nature: it can be found in the rocks of a mountain, in the amplifications of lightning or of a tree, in the edges of a cloud, or in coastlines, and in the delta of a river.

Mathematical fractals are based on equations that are repeated many times according to special computer programs. Look what can be hidden within a sequence of numbers: fractals of extraordinary complexity and beauty, appearing like psychedelic hallucinations (fig. 4.1.a and b)! Why don't we think that *all* the forms of physical bodies can be sequences of relationships and numerical codes? Perhaps bodies are rhythmical pulsations of matter, virtual images of a deeper reality of rhythms and specific sequences.

Let's return to the autopoietic networks that always reproduce

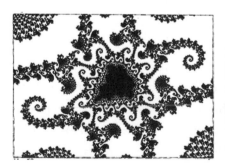 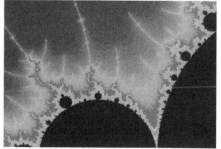

Figure 4.1a and b. Examples of fractals (from *The Beauty of Fractal Images of Complex Dynamical Systems* by H. O. Peitigen and P. H. Richter[6])

themselves: a crystal, a cell, indeed our planet, are subject to autopoietic schemes. Harold Saxton Burr reminds us that the scheme of our body is kept intact, even if new cells replace the old ones. Even if we change, we can always be recognized because the scheme doesn't change and the form remains intact. A child playing with a sand mold of a duck can empty and refill the form with sand many times, but she will keep on reproducing the same little "duck." The code is in the little mold. The cells of our body are multiplying, advancing one after the other on designated paths, directed to the right positions in order to establish the boundaries of the body form. Something similar takes place with the atoms of crystals and the molecules of everything that has form. A virus-infected cell can be told what to do because the virus imposes its own code on the cell, complete with a scheme of "construction." The viral nucleic acid prints proteins, but it is the schema that guides them to the right place to assemble the new virus. The director is not the DNA, but rather patterns of organization. The DNA is the typewriter, but someone else does the typing.

Living systems are governed by patterns and conditioned by their environments, the law of necessity. What the patterns represent is only a part of what is written in what we call the basic code. The basic code is the ensemble of information that allows bodies to exist, the blueprint for the construction. But it is also the "program of the entire evolution," in keeping with Burr's observation that the field of a seed already has the shape of the adult plant. Within the code is written what was represented, what is being represented, what will be represented, and maybe even that which will never have the opportunity to be represented. The basic code is the essence of the thing itself. It is the fractal, which in the film of self-realization gets a glimpse of a different part of itself.

The basic code is revealed a little bit at a time, as it is expressed according to the situation and the need, like a sophisticated computer with which the skill of the operator and the needs of the moment combine to translate some of its many potentialities into action. The becoming *becomes* by expressing only one of many possibilities at a time. If we could read the code of things, the film of the becoming would unfold.

So, the basic codes are a lot more complex than what they express. Water has the same code as ice and vapor, which are different, their differences compelled by need, pressure, and temperature. When we consider the architecture of crystals or beehives, we realize that there are patterns in both the ensembles as well as the individuals. You might think unconscious programs govern the works, but then how would you explain the patterns of the mineral world, where there is no unconscious? Where are the designs that provide the information for the construction of bodies and maintain their identity?

What is the basic code made of? *Informed matter,* an intermediate state between pure matter and combined matter. Pure matter is a still ocean. Each body is like a wave made of combined matter (the crest of the wave) and informed matter (the unseen part under the water, from which the crest derives). While it is not yet possible to measure the code, we can catch some of the interactions and communications in which information is exchanged. And what *is* information? Whatever its nature, information expresses rhythms, vibrations, and disturbances of the field. These spatial areas where the field is disrupted are what we call *informed fields.*

Informed Fields

I won't start by defining information. I will start with an image: the god Hermes, with winged ankles, who runs on the waves of the sea. He is agile, light, fast as lightening. In his hand he carries the golden rod with which he enchants the eyes of humans. A gust of wind, and he is in the cave of Calypso, the goddess who holds Odysseus but who will be forced by Zeus to release him. Homer portrays Hermes as the divine messenger, in one of the first representations of information.

There is a god in information, a god that is the essence of life itself. This god operates everywhere: in science, linguistics, cybernetics, communications, computing, genetics, in all moments of our lives where there are signals that provide elements of knowledge. *To inform* comes from the Latin *in-formare,* meaning "provide a form," to mold according

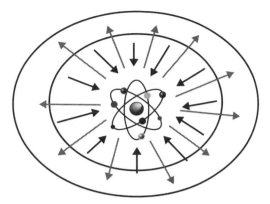

Figure 4.2. The informed field acts on the mass, maintaining its shape, structure, and functions; it also acts on the outside, allowing interactions and communications between fields.

to a shape. Nothing is created without information. Every creator informs the creatures after him (as it says in the Bible, "And He created Man in his image and likeness"). From "intellectually inform" the meaning moves on to "keep informed and give the chance to operate."

Information affects the behavior or form. An example is a driver at a set of traffic lights, or the hormones that change the body of an adolescent. Everything takes place in three phases: the signal must be emitted, then transmitted and received, and finally understood. The object of the signal must be able to receive it (the distracted driver may not see the traffic lights) and to understand it (a wild creature in the forest doesn't understand traffic lights). Everything changes according to the information. As Heraclitus of Ephesus put it, everything flows and nothing is repeated. Even the river is never the same, though it may appear to be so.

Everything is information: the melody that makes you dream or the molecule that activates a receptor. The universe is a net whose knots are the bodies and whose threads are the waves and radiations. The knots may be perceived by the senses, but the threads are not, though they inform, exchange, and organize life.[7] Everything exchanges information with everything in the communication network of the world. Every being and entity sends messages that contain history, conditions, and properties, almost in eagerness to reassert identity and communicate it.

Everything gets repeated this way, to construct the network of information and exchange operating invisibly in the emptiness, in the *apparent void* (fig. 4.2). Everyone knows that our senses are limited, yet we continue to ignore what we cannot perceive. And we lose the faculty to know.

The largest parts of events are not perceivable. Think about how many experiments appear not to work just because we are unable to record the infinitesimal variations produced! In the world of emptiness, things could be different from the world of solids and thus require other paradigms. The history of physics is full of precedents, such as Heisenberg's discovery that space and time behave differently in the infinitively small. Even the vacuum has characteristics, laws, and physiology to investigate.

Although it seems still, matter is in constant motion. Vibrating atoms and dashing particles are confined in narrow spaces. As with a bullet in the barrel of a gun, the narrower the space, the faster the movement. Electrons travel at hundreds of miles per second, protons and neutrons ten times faster. The rotations of protons reach speeds of 3×10^{22}, thirty thousand billion billion times per second. Since electric charges in motion produce magnetic fields, such fields constantly permeate everything. If we travel up in the ladder of dimensions, molecules are not static either, because they are subjected to multiple movements (called stretching, bending, and so on) responsible for vibrations.

"What kind?" you may ask.

"Electromagnetic," you might reply. Correct, but not solely. Take an object, any object, and start to move it. Shake it hard, and think about what is happening in the surrounding air. That moving body is moving small masses of air, and so is adversely affecting its field with elastic types of waves. Never thought about it? The same thing happens in the microscopic world: by vibrating at a very high speed, molecules generate an electromagnetic elastic field because they disturb the nearby environment. Herein lies the secret: electromagnetic waves, which are also elastic.

If the superstrings theory is true, then it fits well with this understanding. The bricks of the world would be microscopic strings, each vibrating in its own way. Every particle of matter and even every media-tor of force would be a string that vibrates, its mode of vibration leaving an imprint in the universe.[8] Think about a violin string: how many pos-sibilities of oscillation—the so-called resonances—how many different types of waves, with peaks and dips equidistant from one another! The string is the same, but different vibrations give rise to different notes. Similarly, the same microscopic strings would be vibrating in differ-ent ways to give the feeling of different masses. The properties of an elementary particle would be determined by the vibration mode of its internal string.[9] Mass is the energy with which the string vibrates, and the "matter" of which the forces and particles are made is always the same. Everything corresponds: the universe is an extraordinary immense symphony!

Whether the theory of strings is true or not, at this point it doesn't matter anymore. Whether they are strings or particles, what differ-ence does it make if there is nothing that keeps matter still? Vibrating strings, vibrating particles, vibrating fields. What appears as a mass is in fact a fluctuating field, rippling matter: the more time passes, the more physicists talk like philosophers again. The uncertainty principle suggests that nothing is ever at rest and everything is subjected to quan-tum agitations. If this were not the case, we would know with absolute precision the position and the velocity of particles, violating the laws of Heisenberg. We can't stop the process of becoming to catch it in its elementary constituents, because that which appears fragmented to the senses in reality is one.

Now we need take care and reason this together. The movements of molecules are autobiographical, individually unique. They are specific. So also are the perturbations in the field. The field thus forms a portrait of the substance, the transcription of its molecules. The vibration in the field is called *fingerprint vibration*. Different substances have different fields, specific to their molecules. For example, the field of hydrogen is specific to hydrogen; it contains features that differentiate it from other substances. If two fields are equal, it means that the two substances are the same; their molecules possess the same chemical and physical structure. Different substances have different fields and emit different signals. The uniqueness is related to the information in the field, which operates according to the given ensemble of molecular signals.

The field created by electromagnetic and elastic fluctuations is *informed;* it contains information about the form, dimensions, weight, and color of the physical substance. It has autonomy, physiology, even potential pathology, qualities, functions, and properties. This applies to the fields of a body's cells as well as to those of objects: all forms of life are governed by codes, whose spatial extent is their field. The field of a medicine, for example, will contain information on the molecular structure, its macroscopic chemical-physical qualities, and its pharmacokinetic medicinal properties. Within the code is written how it works, what it is for, and the action it induces.

The essence of the field is information, which is not solely electromagnetic in nature. While electromagnetism may be the most obvious aspect of the field, it would be a mistake to think that the field is only that. It is one of many properties of a field; to consider it the only one would be like saying, "Look, a walking coat!" while ignoring the man who is wearing it. Information is always a messenger of a beginning or an end or at least a change. It creates, corrects, modifies. It can have a plastic function of regulation or communication. Every entity is governed by information that guides its construction, directs it, and regulates communications. Our contention is that the basic code of things has these three functions:

1. Forming the shape: the instructions of the basic code provide guidelines for the architecture of a body as well as preserving its shape and identity.
2. Organizing, regulating, controlling: the basic code is the designer and the regulator. If the question of why electronic orbits respect only their own specific distances is answered by saying it is due to the forces of attraction and repulsion between the charges, then we have only described the phenomenon without explaining it. This is like saying that a building is shaped a

certain way because there is cement between the bricks holding them together: the architectural design is ignored. The electrical attraction and repulsion (the cement) don't determine the shape; it is the basic code that is responsible.

3. Communicating: bodies communicate by exchanging information between fields through incessant inaudible communications of a very weak intensity, the whisperings of matter.

Let us further explore the evidence that supports the idea that the *codes are responsible for the architecture of matter, its organization and regulation, and its communications and exchanges with the environment.*

5 ⊶ LISTENING TO MOLECULES SING

Molecules sing, tell stories, and convey intentions. And listening to them creates a fascination that leads to alignment with the singing. This has been happening for years in my laboratory, where we listen to the singing of molecules. Then we watch the effect of their singing on cells that—like Ulysses encountering the sirens—are spellbound by the sound. Then, as if they were following a magical piper, they move in the direction of the song. Please forgive me if I write poetically about my discovery, but I do not think that seriousness increases the value of research, and I do believe in the *magic* of nature.

Unlike pure matter, particles, atoms, and molecules are dancing. As the microscopic spheres move they release elastic waves that in turn move the air and perturb the space around them. Think again about a violin string: the moving of air caused by the vibration is perceived as a sound. The area around the string is not the same as before it vibrates because when the string moves, even the space vibrates and dances to that rhythm. There is a story about a man who became a god, Orpheus. His voice, accompanied by the lyre, charmed all he met: men, women, animals, plants, even mountains. The beasts became tame at his feet, the rocks groaned with emotion. He made, as the poet Rilke wrote, "more sorrow pour out of one lyre than from all sorrowing women." His enchanting music wrapped "forest and lake, street and village, field, river, and beasts" in a world of sorrow, around which a crying sun "and star-filled silent sky revolved, a crying sky with deformed stars."[1] Myth teaches—and in myths there is always truth—that modulated sound can transform matter: this is the meaning of enchanting. This chapter is the story of an enchantment.

Objects Also Communicate

The exchange of information is life, and there is no life without communication or communication in the absence of life. The horse communicates by neighing, the dog by barking, and so on, but animals can also communicate in other ways: gestures, postures, and expressions. Bees communicate by dancing along invisible routes in the air with precise geometry. There are also extrasensory communications involving plants and animals, and even objects communicate with each other and with the environment, thanks to their informed fields.

These may be imperceptible changes, known so far only through the intuitions of some philosophers and physicists, but research must be right here, at the point where there are no explanations, because everything is inexplicable when it is new. If someone in the Middle Ages had tried to communicate through a radio, he would have been burned at the stake. So, without prejudice, let's try to imagine objects that communicate. Things may seem to be quiet, because this is how we expect them to be, silent and passive. However, let's find the courage to visualize something different, to try to observe from a different point of view.

If we free our imagination and envisage putting a piece of iron close to a wooden surface, though the senses register nothing, at the level of the fields, there would be some exchanges where the iron informs the wood and vice versa. The iron shares what it means to be iron, communicates how it is made, its qualities, its physics and chemistry, its properties and potentials. It shares its basic code. The wood does the same by communicating about itself to the iron. During the exchange their molecular structures appear unaltered, but the communications are registered in the field. They will stay there for an instant or maybe forever, like the dog that has sniffed a person will still be able to recognize that person in the future. Just as we have seen about water (in chapter 3), which can receive and emit signals by entering a coherent state, all types of matter can emit signals rich in information and as powerful as laser beams.

What is the information? It is energy being transmitted on electromagnetic waves, propagating variations in the field. It is the energy that moves, but not the matter. How can we imagine it? Like the waves generated by a breeze blowing across a wheat field, as described by Leonardo da Vinci: "There are many times when the wave flees the place of its creation, while the water does not; like the waves made in a field of grain by the wind, where we see the waves running across the field while the grain remains in place."

The work of Sir Jagadish Chandra Bose done in India during the end of the nineteenth and the beginning of the twentieth century confirmed that things communicate, even if on such a scale as to be difficult to notice. He studied electric waves and the physiology of plants. A year before Marconi's discovery, he transmitted radio waves from the conference room of the Calcutta Town Hall, through three walls, to a room twenty-two yards away. Three years later he published his own observations on the behavior of the electric waves in magazines such as *Nature* and *Proceedings of the Royal Society*.

The following year, Bose noticed that, with constant use, the sensitivity of his detector mechanism became reduced and that it would return to normal after a period of rest. This "strange fact" led him to the conclusion that metals, just like humans and animals, may need to recover from strain and that the dividing line between *not living* metals and *living* organisms was very tenuous.[2] He began a comparative study of the curves of molecular reactions in inorganic substances and those in living animal tissue. Bose observed that the graphs of the oxidation of warmed-up magnetic iron looked like those of the muscles: response and recovery diminished with exertion, and gentle massages or tepid water baths could eliminate fatigue. The metals seemed to react in a way similar to animals. He noticed that on a metallic surface that had been corroded by acid, then polished so well that every trace of the corrosion was gone, the points previously attacked had different reactions than the other areas. Bose attributed those reactions to the prolonging of the memory of the corrosion.

He asked the secretary of the Royal Society to verify his experiments by submitting readouts of response curves. The secretary duly noted that there was nothing extraordinary in the curves he observed. When Bose asked him what he thought the curves represented, he responded, "Muscle reactions." When Bose proved that they were the response curves of a tin can, Sir Michael Foster organized a conference that was held on May 10, 1901, at the Royal Institution. It ended with the following words from Bose:

I have shown you this evening autographic records of history of stress and strain in the living and non-living. How similar are the writings! So similar indeed that you cannot tell one apart from the other. Among such phenomena, how can we draw a line of demarcation and say, here the physical ends and there the physiological begins? Such absolute barriers do not exist.

It was when I came upon the mute witness of these self-made records, and perceived in them one phase of a pervading unity that bears within it all things—the mote that quivers in ripples of light, the teeming life upon our earth, and the radiant suns that shine above us—it was then that I understood for the first time a little of that message proclaimed by our ancestors on the banks of the Ganges thirty centuries ago: "They who see but one, in all the changing manifoldness of this universe, unto them belongs the Eternal Truth—unto none else, unto none else!"[3]

After the success at the Royal Society, which accepted his data unanimously, Bose discovered other analogies between animal and plant physiology. Bose discovered that, like metals and muscles, plants also react to chemical attacks and that they can be anesthetized, such as with chloroform. The death of a plant is accompanied by spasms as in animals, and in the moment of death, there is an enormous electric force. In chapter 10, we will see how plants perceive the death of other beings. The death of five hundred fresh peas can develop enough volts to electrocute whoever is cooking them, except that peas are rarely connected in series. George Bernard Shaw, a vegetarian who was also against dissection, was shaken when he witnessed in Bose's laboratory, through one of Bose's magnifiers, a cabbage leaf scalded to death; he dedicated his collection of works to Bose, with the words, "From the least to the greatest living biologist."

Bose, inventor of the wireless telegraph, refused to patent his discoveries, wanting them to remain in the public domain, to give them to humanity. In 1917 he founded the Institute of Research in Calcutta and designed an instrument that he called a "crescograph," which was able to record every movement of a plant at magnifications of up to ten thousand times. It could record the rate of growth of the plants. He demonstrated that plant growth proceeded with rhythmic impulses, every one of them showing a quick increase followed by a partial regression, equal to one-quarter of the distance gained. Furthermore, growth could be slowed down or stopped by simply touching the plant; some plants, if mistreated, were more stimulated to grow. With the crescograph, capable of revealing infinitesimal variations in the rate of growth (1/1,500 millionths of an inch every second), it was possible over the period of only one-quarter of an hour to determine the action of fertilizers, nutrients, or electric currents given to a plant.

In 1920, after harsh criticism and ideological battles, the University of London accepted the results. The subsequent publication of his book, *The Physiology of the Ascent of Sap,* enchanted, among others, the philosopher Henri Bergson. In his last years Bose was appointed a member of the Commission for Intercultural Collaboration of the League of Nations, together with Einstein, Lorentz, and Murray. Just before he died he said:

> In my research on the action of forces on matter, I was amazed to find boundary lines vanishing and to discover points of contact emerging between the Living and non-Living. My first work in the

region of invisible lights made me realize how, in the midst of a luminous ocean, we stood almost blind. Just as in following light from visible to invisible our range of investigation transcends our physical sight, so also the problem of the great mystery of Life and Death is brought a little nearer solution, when, in the realm of the Living we pass from the Voiced to the Unvoiced.[4]

Medicines that Speak to People

Far more recently, certain disciplines of medicine, such as electro-acupuncture, applied kinesiology, and auricular medicine, have been exploring ways to assess whether something has changed and according to which parameters, such as with substances that induce changes in matter, like medicines. If a transmission takes place between a medicine and a person via a simple contact, we find biological variations. But could it be enough to put medicine in contact with a human being to create communication? Let's see what has been discovered.

For a couple of centuries in the West (and at least five thousand years in the East) it has been known that there are lines of preferential flows of energies of the organs in the human body. These channels, called *meridians,* run through the body internally, and on certain surface areas of the skin there are *points* through which it is possible to intervene and access these meridians. In the 1950s the German doctor Reinhard Voll, an acupuncturist who was also an expert in radio transmissions, developed a method to locate and measure acupuncture points and then through electric stimulations to change the potential. This method, known as electro-acupuncture, uses instruments that measure the conductibility of the skin. Over the acupuncture points this is usually higher. The patient is connected to the machinery through a metal electrode that acts as a negative pole. With the other electrode (positive pole) the doctor tests the points, referring to each organ and its functions (fig. 5.1). The resulting values provide information about the organ's energy level and health status. For example, if the points of the Stomach meridian indicate high figures, there may be gastric inflam-

Figure 5.1. Scheme of the measuring of points through electro-acupuncture machinery (reproduced with the kind permission of Erich Rasche)

mation, while low figures for the empathic cells may indicate some liver degeneration.

Voll discovered the *testing of medication* by chance, during a demonstration of his system. He measured the points of a colleague whose values changed noticeably after a short break between testing of only fifteen minutes. "What did you do during the fifteen minute break?" asked Voll of the other doctor. "I only took a medicine that I take every day," was the answer. Voll must have had an unexpected intuition, so he said, "Tomorrow bring me a sample of your medicine and we will repeat the measurement before and after you take it." The following day, Voll demonstrated that even by simply holding the medicine, the values of the connecting points were changed. The medicine test was born.

Voll had found a way to assess the compatibility between a patient and a medicine and thus a way to choose the most suitable medicine for a given person.[5] This was done by including a metal vessel containing the specific medicine that was to be tested for that particular patient in a circuit formed by the patient and the machinery. If the inclusion of the medicine improved the value previously recorded, it was a sign of compatibility: that medicine was good for that person. This test is possible because of the *resonance* principle. In physics, two systems resonate with each other when their oscillations vibrate at the same frequency. There is resonance whenever a vibrating tuning fork starts up; resonance is responsible when the voice of a singer cracks a glass.

There are other techniques in medicine to test drugs. In applied kinesiology muscular tests are used. The patient must push a muscle (usually the deltoid or the quadriceps) against a resistance, such as the doctor's hand. The strength is measured. Then it is measured again after there has been contact between the patient and the sample of the medicine being tested. If the strength doesn't change or increases, it is a sign of good compatibility; if it decreases, the remedy is not suitable.[6]

Even in the auricular medicine invented by the French doctor Nogier, remedies can be tested for vibration on the subject. The parameter is the amplitude of the pressure measured at the wrist. The compatibility of a medicine put in contact with the surface of the ear is evaluated in terms of an increase or decrease in the force of arterial blood pressure (known as ACR: *auricular-cardiac reflex*), infinitesimal variations that an expert hand can detect.[7]

In different ways, the three methods reveal resonance between the medicine and the subject. They all indicate that fields are capable of interacting not only with the patient but also *with the medicine*. There can be no resonance if one of the two parties does not transmit: therefore, a medicine capable of emitting *something* is informing the patient about itself through mere contact. Remember the hypothesis that bodies interact by exchanging through their fields?

A particular resonance between human bodies—a "tuning fork effect" between the fields of people—was also observed by Nogier. While the French doctor was testing the ACR on a patient's wrist, not far from him, his students perceived on their own wrists the variations that Nogier perceived on that of the patient.[8]

In the 1980s many researchers observed that homeopathic remedies emit frequencies capable of interfering with the body (the electrocardiograph and the electroencephalograph are examples of ways of registering and measuring changes in electric activity levels in the body). This property was only talked about in the context of homeopathic and not chemical remedies because the research in the three disciplines mentioned above had focused on natural and homeopathic treatments.

Studies done by German doctors suggested that the radiation of these remedies is electromagnetic in nature. Fritz Kramer, one of the founders of electro-acupuncture, was successful in testing medicines not placed in a little metal container, but at a minimum distance. The signals from the homeopathic remedy were shown to interfere at distances of more than 5 mm. The range involved provoked Kramer to think he may be dealing with waves similar to radio waves.[9]

Kramer's insights were confirmed by another German doctor, Franz Morell, who used a transmitter to test homeopathic remedies in the late 1970s. The transmitting part of the apparatus is placed on the medicines, while the aerial that receives the signals was connected to an electro-acupuncture circuit. Using this method, Morell could test groups of medicines at a distance. It followed that the homeopathic remedies emitted radio waves that could affect the subjects to which they were transmitted via cable or via wireless means.[10]

The person who first realized that the properties of a homeopathic remedy could be transferred passively to water was once again a German, H. Schimmel, a dentist. First he used the electro-acupuncture procedure to determine that a vial of a certain homeopathic remedy was compatible for a certain patient. Then he would put the vial in a glass of water (which had shown itself to be inert in relation to the patient) and leave it for a while. After some time he would collect a vial of water from the glass and test it on the patient. Even though the remedy had not combined with the water, the water produced the same test results as the medicine. This suggested that some therapeutic radiation had passed from the solution contained in the vial to the water. We might add that the frequencies of the remedy, passing through the glass vial, transmitted their movement to the water, which consequently started to vibrate with those frequencies, thus becoming a *copy* of the remedy. The new vial was, however, inactive when put into boiling water because the boiling destroys the radiation emitted by the homeopathic remedies.[11]

Years later the Austrian biologist Christian Endler, not knowing about Schimmel's work, discovered that homeopathic remedies react not only via contact but also by spreading frequencies in the environment. Endler liked to experiment on tadpoles. By dispersing a homeopathic thyroid hormone (thyroxin) in the tadpoles' water he discovered that it influences their speed of metamorphosis. Then in the nineties Endler did a test in which he sealed solutions of homeopathic thyroxin (dilutions DH30) in glass vials and immersed them for weeks in the tanks containing tadpoles. He obtained effects similar to those he had previously obtained. Endler then concluded that homeopathic remedies could emit certain frequencies able to pass through glass,[12] just as had been observed for years in electro-acupuncture practice.

The great tapestry of the transfer of medicines was taking shape, woven by the many researchers producing similar results. But nobody had yet put together all the data that had been amassed. It was as if everyone were weaving the same fabric in the dark. If a medicine can transfer information to a person, could it also cure that person by the same principle? Could medical properties be transferred without the person actually having to take the medicine? Should notions such as this be considered science fiction or the dawn of a new physics?

Curing with Waves?

It was Morell who took the next step of making electro-acupuncture not only diagnostic but also therapeutic. In order to capture the waves of homeopathic remedies, modulate them in a carrier frequency, and then transmit them to a patient, he invented an apparatus that is similar to an amplifier with high impedance; it has an entry in which the remedy is placed and an output connected to the patient. It demonstrates that the oscillations of the remedy are transferred and that they modify the values measured at specific points. Later on the output of the circuit was inserted into a small glass bottle full of water to see if the therapeutic messages would saturate into it. The results confirmed that this had indeed occurred. Morell had found a way to *duplicate* homeopathic remedies.

In the late 1980s many German and Italian electro-acupuncturists started to *load* small bottles of water with the frequencies emitted by the remedies, preparing medicines of this irradiated water loaded with the homeopathic information for their patients. Compared with homeopathy there was no dilution, but rather frequency/information input from the outside. The water became a photocopy of the remedy, confirming four significant things: (1) that information can be received in this way, (2) that remedies emit waves, (3) that these can go through glass, and (4) that homeopathy thus has nothing to do with molecules because the waves are emitted from matter and imprinted in the water by successions, making the traditional grounds of objection used by homeopathy's detractors irrelevant.

Until 1989 it was thought that the transfers were possible only with homeopathic remedies since, compared with synthetic ones, they emit waves.[13] But since it is also possible to test food and other substances for resonance, we came to the conclusion that it was plausible that everything transmits information and that the frequencies of synthetic medicines are also useful for curing. We understood that if this indeed were the case, it might be possible to show something that until that point had seemed unthinkable, that matter *emits signals containing information.* Remember that chemistry talks about fingerprint vibrations! The question our research team wanted to answer was whether all of this is valid just for homeopathic remedies, or could the information from traditional medicines also be transferred?

In biochemistry, it is thought that the action of a conventionally administered medicine on a cell is like a key opening the lock of a door. The lock is the cell's receptor membrane. When the right key (the medicine's molecule) enters the lock, the lock can be clicked open and a reaction will be triggered. But it is not the only way to act on the lock: it is possible to open a door even without a key if you have a remote control that sends the right frequency, which, coming into resonance with the lock, clicks it as if with a key. Probably the interaction between signal and cell comes as shown on page 66 (fig. 5.2). We decided to investigate what happens when, instead of a medicine, we send its signal to the cell: we wanted to know if the door would open and the same pharmacological action would take place as when the medicine itself is used. It would be as if the *voice* were substituted for the medicine; instead of administering the medicine in the traditional manner, waves would be transmitted. If all the information necessary to heal were in what the medicine emits, then there would be no need to administer the actual mass of the medicine to the human body.

What we have found is that the result is the same whether the molecules of the medicine physically make contact or send signals to the cells, since the chemistry and physics of the molecular signals follow parallel paths. The result is medicine that is not chemical anymore, but informational. The cells listen to the singing, understand it, and know how to respond appropriately. Things communicate with things.[14] In order to explain how we were able to do this, we need to take another close look into the nature of matter. As we have seen, observable or combined matter is always in motion. It becomes immobilized in only one case: at absolute zero, the lowest possible temperature, 0° Kelvin (273°C below zero). In this frozen state, the molecular structure reaches the highest possible order, since the molecules suspend any oscillation, caught in crystal lattices, such as sculpture in ice.

Now imagine progressively providing thermal energy to such frozen

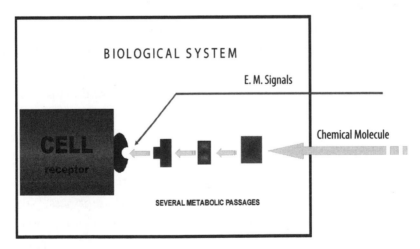

Figure 5.2. Hypothesis regarding the cellular mechanism for receiving pharmacological signals. Physical and chemical signals could act on the same receptors.

matter. As they warm up, atoms and molecules start to vibrate again, not with their harmonic rhythms, but with chaotic oscillations (think about a pan of boiling water). Heat is responsible for chaos only until the molecules start to vibrate at their own rhythm, which is the "identity level" of that substance: at that moment, its *music* starts. Heat is essential to life because a system at absolute zero seems dead, with life suspended. From an order with no life, through an apparent disorder, we reach the order of life. The physiological frequencies of bodies lie at precise levels of temperature above absolute zero. At these frequencies they will emit information; they express their codes.

Let's imagine getting microscopically small to enter the fantastic scenery of the atomic world. At absolute zero gigantic crystal spheres are frozen in an ice landscape in the most spectral silence. Then, awakened by heat, they start to move again with their rhythms. They rotate like one of Dante's circles, their rhythms producing frightening roars. They are the *voices* of the molecules. Their vibrations do not emit a melodic music; they sound rather like cries from fantastic beings: whistles, sirens, blasts, shrills, all in a rhythm that is repeated relentlessly. What is more disturbing to us as we explore is the very strong wind blowing in every direction. We must proceed as if we were crawling, almost holding on to the floor.

Why the wind, and where does it come from? It comes from the tremendous molecular motion of the unbelievable masses, which in our macroscopic reality is known as *field perturbation*. Don't forget that any movement always creates motion around it. Maybe we don't notice, but while we walk (now I mean in our dimension) we move tremendous amounts of air. A dog that passes by, the train that zips by, leaving a breeze—all that exists moves and disturbs the environment. Think (if we could only see them!) what frightening currents are generated in our cities: thousands of movements move tons of "emptiness," streams crisscrossing and intertwining. This happens also with things that look immobile. A stone, an object, a building appear as if they do not move at all. But indeed they do.

Nothing is still, except in the abyss of absolute zero. The surfaces of water are not static, but subjected to microscopic movements, like the rest of condensed matter, from macro to micro. The table on which I write expresses imperceptible movements that in turn move something in the environment and signal its presence. Someone more sensitive could feel this. Think about blind people who feel something silent is coming close. They are sensing the elastic waves transmitted by what is moving. Interestingly, the same thing happens also with still objects: in walking, a blind person feels the presence of an obstacle before hitting it. Although apparently still, the object deforms the space around it, moving quantities of air in waves that propagate centrifugally. Although most of us would need to touch the object in order to sense it with our eyes closed, a blind person (or other person with heightened sensitivity) can feel the waves spreading from what for her is an obstacle and can stop herself in time. (It is more likely for a blind person to fall into a hole in front of her because the hole doesn't move anything.) This is resonance between fields. The field of the blind person interacts with that of an object, which communicates its presence.

Everything vibrates and emits elastic waves that can be translated into sound, so everything plays music with those little bells that are molecules. The molecular movements are not chaotic, improvised by chance, but governed by physical laws and mathematical rules: they are ordered oscillations. Just as music is an expression of harmonic mathematical rules, so the molecular movements are rhythmic sequences of sound, the music of microscopic spheres. Information is in the rhythm of the molecules: variations, alternation, and pulsation. Things are different because the molecules are different (in quality, number, and disposition), and therefore they express different music. Each substance has its own sound.

Researchers realized that if these sounds could be rendered audible to the human ear, we could listen to the noise of matter: the *voice* of a plant, of a stone, of any object: a can, a cell, a protein, or a medicine. In the Czech Republic, Josef Havel, professor of chemistry at Brno University, has recorded the voice of a protein. To each molecular motion of the protein, separated by capillary electrophoresis, Havel associated a note and he could play his melodic music for that protein while it oriented in an electric field.[15] In 1999, Cyril Smith, a Stanford professor of physics, played the (cacophonic) music of thyroxin during a convention, but the performance was immediately suspended because many people in the audience suddenly experienced tachycardia.[16] What happened? The music had reproduced one of the effects of an excess of thyroid hormone: thyroxin overdose.

We have discovered that everything that exists emits a *voice,* which in turn contains resonant sequences. It is as if an invisible heart beats in every body, pulsating with a rhythm that is unique; it is these rhythms that determine the specificity of bodies. This leads to the amazing idea that if we could modify the codes, maybe we could change the structure of the bodies! Let me now invite you next to peek inside our laboratory and see how we have developed a method of recording and transferring a medicine's basic code.

6 ○─● THE POWER OF TFF

A Radio That Transmits Medicine

Even though we still don't have the basic codes of all the medicines, we have been able to record them and transfer them to reproduce the effects. We start with a sample of active principle without exipients (pharmacologically inactive substances). The method of transfer is called TFF (*Trasferimento Farmacologico Frequenziale*, i.e., *Transfer Pharmacological Frequency*). The scheme is simple: at the entry of the apparatus we have devised there is an electrode plate of brass on which the substance to be transferred is placed, enclosed in a glass vial. The field perturbation that bears the signature of the medicine is transmitted to the brass, and the signal enters the unit. An identical electrode connected to the output is placed on a small glass bottle containing purified water and a saline or hydro-alcoholic solution (fig. 6.1). The information that the medicine emits in the form of waves is captured by the apparatus, amplified,* and then transmitted through the glass into the liquid, as described in the chapter dedicated to water (chapter 3). The TFF imprints all the properties of the medicine, including the curative ones, into the water sample, which conserves them for later administration to an organism.

---

*For the first few years of experimentation, we used an electro-acupuncture apparatus (MORA III 200), then others, until reaching a prototype of our own construction.

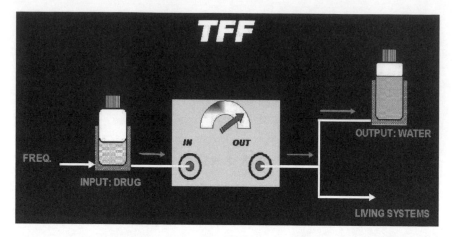

Figure 6.1. Transfer Pharmacological Frequency with two possible outputs, one directly to a patient and the other to water, which is then administered to a patient.

Although remaining chemically water, from a physical point of view, the solutions are a copy of the starting medicine. The relationship between water and water treated with TFF is the same that exists between water and ice: only the physical structure differentiates what from a chemical perspective is simply H_2O. They are not the same thing, however: think about the difference between bathing in water or in ice!

We have found it is also possible to do the transfer without water, to transfer the properties of a medicine directly to an organism via a cable connected to the output of the apparatus. A direct transmission by contact has the same effects as when water is used as a carrier; in fact, it is quicker than with the chemical medicine because the subject is in direct contact with the waves. In addition to liquid, transmission is also possible via solids (such as metallic alloys) or informational carriers.

Let's go back to the music of the medicines and how the transfer takes place. Like tiny radio stations, medicine molecules vibrate at their own frequency, emitting information of their own identity, which reproduces their pharmacological action. As the initial signals are extremely weak, the medicine needs to be stimulated to reach sufficient power by using a particular frequency emitted by a generator. The amplifier circuit needs to have high impedance, with bandwidth for low frequencies (from 4 Hz to 35 kHz: from the infrasound to the ultrasound, including the range of hearing), powered by batteries. However, the device is not just a frequency amplifier; with an electrode input and output serving as antennas, it actually behaves like a *radio*. It is no accident that its frequency band includes the spectrum of the audible, ranging from some infrasound around 4 Hz, the full range of audible sounds to the human ear (from about 16 Hz to 20 Hz), up to several ultrasounds (around 35 kHz).

Here is how the TFF proves possible from a physical point of view. When the molecules of the medicine receive the wave from the generator, they enter a state of strong excitement: the oscillations become more amplified, creating stronger emission signals (like in a radio). The input electrode picks up this "concert" of incoming waves, just as around a radio's antenna there are electrostatic field variations (voltage differences). The apparatus records these easily by having a very high impedance, converting the sound signals to electrical fields that are amplified properly. They are then transported to undergo the opposite process of transmission, just like in the radio. What is heard is variations of electrostatic fields, sound waves; the information is the frequency by which they vary.

Now let's try to imagine what can happen to the water that receives the output signals. If we were to get very small once again and enter the microscopic world of water, we could hear a slight, continuous background sound that is the vibration of the water. Imagine being there when suddenly there is a wave of loud music, but harmonically organized, as if the monotonous vibration of a violin string had been replaced by an entire orchestral symphony of strings.

The vibration of the water is overwhelmed by the incoming power, and the molecules start to oscillate with new rhythms; in a short time all the water takes on the frequencies of the medicine. From a mechanical point of view, once sound waves are received, the molecules start to transmit these signals and continue to bump into each other with that rhythm. When the shock waves reach the walls of the small bottle, it, being made of glass, partly allows them to be passed through (in fact, that small bottle can be tested in many ways, and we see that it too emits frequencies) and partly reflects the waves back into itself, like a mirror. This allows the signals to trigger a new set of mechanical shocks that spread in the whole bottle until the walls reflect them again; the pattern is perpetuated, self-sustained over time.

The preparation of homeopathic solutions may similarly trigger shock waves that convey the music of a substance to the water molecules. In TFF the process of shock waves is started by the power of the apparatus, whereas in homeopathy it is done through dynamization or succussion. In that case also we can envision waves being reflected from a container, as up until the eighteenth century the containers would have been made of glass (plastic didn't exist). Right from the start of TFF, glass has played a fundamental role. In fact, TFF of water in containers of different materials—like polyethylene or Pyrex of a great thickness, which absorb mechanical waves—doesn't work and the water remains inactive. This proves that in the informed fields there is also an important mechanical component, probably acoustic.

This would help explain why the increased conductibility values and the excess heat in homeopathic remedies and in water activated with TFF would continue to increase with time, as has been shown by measurements made after three months or even three years. It could mean that the molecular dance of glass—the source of its reflecting property—continues its rhythm without ever dying out, an energy that is self-preserving and autoregenerating.

The action of the glass on the mechanical waves also may explain what Vittorio Elia has called the "volume effect," a particular phenomenon that can be observed by measuring the conductibility of homeopathic remedies and TFF.[1] If a solution is divided by pouring it into three containers, each one of a differing volume, the result is that, contrary to any chemical logic, the smaller the volume, the greater the conductibility. Reducing the volume proportionally increases the surface, so there is more contact between water and glass, which reflects a greater number of waves.

This leads to the conclusion that water activated with TFF and homeopathic solutions both behave in an autocatalytic way, as dissipative structures. Let's see what is at work. Heating a pot of water to boiling creates a state transition from water to steam; if we cool it down, it will change from water to ice. The disorder increases and decreases by heating and cooling, but at a certain point, a thermal balance is established in which the bodies all have the same environmental temperature. At this point the codes best express their information. If only heat is exchanged, it is described as a "closed system." An "open system," on the other hand, exchanges *nonchaotic* energy (matter, electromagnetic signals, information) with the environment, and the opposite happens than with a closed system: when heat is provided, the order increases.

Every input of energy transmits information, says German physicist Fritz Albert Popp, "no system can be excited so chaotically that among its particles will not occur, at least for brief periods in confined spaces, correspondences and correlations, such as when two particles vibrate almost synchronously with each other."[2] In closed systems (the boiling water in the pot) the regular oscillations are very brief and infrequent, while in open systems the energy always transfers consistent information, producing ordered structure. In open systems (living systems) the supply of energy produces "restructuring of matter, new rhythms in the movements of particles. It also produces: gradients, fluctuations, stable space structures, oscillations, images, and shapes."[3]

Ilya Prigogine was the first to observe that some systems spontaneously form complex, sometimes chaotic, structures when they dissipate, that is, disperse incoming energy evenly throughout the system. His model of a "dissipative system" as an open system operating out of thermodynamic equilibrium in an environment with which it exchanges energy and matter won him the Nobel Prize for Chemistry in 1977. Dissipative structures have what can be regarded as thermodynamically steady states; they are *self-catalytic* because they tend to remain in time.

Activated with TFF or with simple homeopathic succussion, water reacts to the waves released from the medicine or from the water itself by becoming organized in an autocatalytic way, with coherent oscillations with a long-range action, typical of systems that are able to organize through information. Calorimetric and conductivity studies provide experimental confirmation that the dilution of homeopathic remedies in water brings about new and unforeseen changes.[4] That is to say, the parameters (calorimetric and water conductibility) vary, attaining maximum values when the remedy is diluted with double or triple the volume of water.[5] This confirms that the homeopathic dilutions are in the paradigm of the dissipative structures (far from thermodynamic balance) recognized and accepted by so-called official science. Nothing is heretical or nonscientific in the physics of homeopathic remedies.

The First Results with TFF

As we have seen, using TFF is like recording the singing of the molecules on the "magnetic recording tape" of water. Once recorded, the medicinal sound can then be reproduced. The next step was to confirm that when the signal is carried by water the recipient cells would recognize it as a medicine, which we could assess by noting whether it had the same therapeutic affect. The first experiments with electronic transfers of chemical drugs began in the late 1980s, and the first results were reached in May of 1990 by my research group, which would later become the Alberto Sorti Institute of Research (IDRAS) of Turin. Having used TFF to impregnate water with information pertaining to the basic code of a particular medicine, we then could give solutions orally (sublingually) or parenterally (under the skin, intramuscularly, or intravenously).

The name of the first treated patient was Rubens. He was a four-year-old Norwegian cat with long fur, with a grave form of gastroenteritis caused by a feline parasite, the *Haemobartonella*, that had almost completely destroyed his white and red blood cells. He was so debilitated that he weighed half his normal weight. Struck by high fever, he was fighting for his life. After a homeopathic treatment that proved futile, we attempted a (TFF) transfer of tetracycline, with a dose of ten drops every four hours, day and night. After just two days of therapy, the cat was showing improvement in sensorial and general conditions; by the sixth day blood counts were almost back to normal, and the serum-diagnosis of the parasite was less positive. After another two weeks he was clinically cured, his weight was returning to normal, and his condition was generally optimal. After two months the serum-diagnosis was negative, and the cat was confirmed fully cured, and there weren't any relapses; ten years later Rubens was still alive. The TFF transfer of tetracycline gave the same results as the antibiotic, with more incisive and definitive results in which the serum values in the blood were zeroed to normal, something that rarely happens, even with medicine. Waves cured the cat just as well as, if not better than, molecules. There were another two similar cases, also treated with the TFF of tetracycline and thiacetarsamide, and they were cured just as well. The theory was confirmed: molecules can sing.

At that point experimentation with humans started. A man suffering from an acute ear infection was cured after a week of intake, administered every four hours, of TFF of another two antibiotics, amoxicillin and dicloxacillin. We obtained fantastic results in the areas of pharyngitis, tracheitis, ear infection, arthritis, migraines, urethritis, and even a case of whooping cough in an adult who was cured after a few days of miocamycin TFF.[6]

We also began to transfer pharmacological signals directly, with the patient connected to the circuit through a cable and metallic electrode. This enables us to obtain answers immediately during the transfer, even faster than with the molecular medicine. This way the TFF of some anti-inflammatories (bromeline, deoxyribonuclease, and nimesulide) had sedated the pain in acute pharyngitis by a good 80 percent even during the transfer, but in the following days nothing could be done for the remaining pain and the cough that occurred in the meantime. It was the beginning of the whooping cough, which then was successfully treated with the appropriate antibiotic via the TFF. An important thing to note here is that if the TFF had been working as a placebo—possible in any experiments with medicines on humans—it would have been effective against the whooping cough as well as the pain, even though the medicines and their *copy* in the water weren't useful. In fact TFF doesn't work when the medicine it is copied from is not effective in the given instance. We tried more cases with the TFF procedure to transfer anxiolytics (benzo-diazepine, mostly diazepam and bromazepam); the cases where the TFF provided positive results numbered twenty-seven out of thirty-two.[7]

A year later, we had amassed eighty-six cases that had responded therapeutically to TFF treatment.[8] The most interesting were cases of heroin addiction. Every day for six days, phials of TFF of heroin were injected into the patient's vein; this counteracted the withdrawal symptoms, causing them to cease within fifteen to twenty minutes after intake of the TFF. The heroin information transmitted into the physiology of the patient's body was "exchanged" for heroin, thus reversing the deficiency symptoms. It seems significant that the TFF treatment was administered in the same way that the drug would have been administered: taken orally, heroin has almost no effect.[9] We followed five more cases of dependency where TFF treatments helped to reduce the withdrawal symptoms, without euphoric effect, for reasons that will be explained later on.

Other cases treated successfully involved menopause symptoms (hot flashes, sweating, tachycardia), with TFF estrogens or progestogens; allergic reactions, with TFF of antihistamine;[10] and eighty-nine cases of inflammatory and painful diseases, with TFF of different anti-inflammatory cures , seventy-four percent of which achieved positive results.[11]

We also treated Parkinson's disease, with TFF of dopamine. The disease generally improves with the intake of this substance, which is missing in the affected areas of the brain. Since dopamine can't reach the brain, one of its precursors is given, L-DOPA, which in the nervous system is transformed into dopamine. Unfortunately, this transformation can also happen in other parts of the body, with numerous bad side effects. These problems made a forty-seven-year-old woman suspend treatment with the L-DOPA, as the side effects were even worse for her than the disease itself. Without the L-DOPA, she was suffering from muscle stiffness and would lose her balance and frequently fall. She lost many automatisms and acquired a clumsiness that stopped her from getting into a car or getting the key into a lock. Her hands shook, her voice was trembling, salivation and sweating had increased; she suffered from memory loss, with anxiety and depression. She agreed to experiment with TFF, the first time this approach had been tried with this disease. We did not use a TFF of L-DOPA (that would have then been converted to dopamine). Instead it was a TFF of pure dopamine, as we thought it plausible that the signals could spread to the brain, overcoming any anatomic barrier.

After a week of treatment (ten sublingual drops four times a day), her loss of voice disappeared (this hadn't even improved with the previous L-DOPA therapy), and her clumsiness, slowness, and ability to lift her foot had all improved. After a month, the patient didn't have balance problems anymore, or memory loss, or depression, and her walking had improved. "It is as if the effects were accumulating. Everybody says that it's like a miracle," she commented. However the tremor in her hand was still there. The nonresponse (or perhaps delayed response) regarding the tremor is typical of dopamine. In this case also, if the results were due to placebo, all the symptoms would have disappeared, not only those usually derived from the medicine. The improvement continued. After two months she started walking almost perfectly with an absence of any symptom, excluding the tremor. The sense of improved well-being persisted in the following years, during which she continued the intake of TFF, three or four times a day.

Other cases of Parkinson's were treated with TFF dopamine, whose therapeutic effects were manifested always within three weeks, and sometimes after as little as ten to fifteen days. In the case of an elderly woman who hadn't been taking medicine for a year because she had a psychiatric condition with disorientation and hallucinations, the TFF improved her condition, allowing the patient, in less than a month, to walk normally as well as to regress the psychiatric symptoms so she could partially get back to her life. It appears that in this case the TFF of dopamine could have cleaned her field of the side effects of the L-DOPA taken previously. With the TFF we noticed that the symptoms regress at the same pace and in the same order in which they start again if therapy is discontinued.

In cases involving Parkinson's, where TFF was added to the pre-existing polytherapy, the patients, in less than a month, were able to get out of bed and dress by themselves once again. In other cases already under drug treatment, the addition of TFF prolonged the effect of the molecular chemistry: when taken in between drug treatments, the action of the TFF helps the patient to develop fewer or no symptoms in the interval between treatments; some patients were able to reduce their pharmacological therapy instead of increasing it.

Another type of application is TFF of thyroxin in patients with hypothyroidism, treated only with TFF or TFF and pharmacological therapy together (at a low dosage). The body's response has always been excellent, with an absence of the normal symptoms and improvement or normalization of the hematic parameters.

Laboratory Experiments

A survey was done of experimental data with TFF tested on biological models of various types: isolated organs, unicellular organisms, and animal and plant models. We documented that, like histamine itself, the TFF of histamine has a dilating action in the coronary outflow of a guinea pig heart (fig. 6.2).[12] We tested this in an experiment carried out

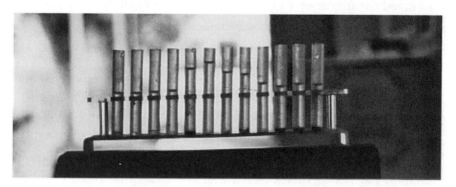

Figure 6.2. July 10, 1992, at the U200 of INSERM in Paris (Jacques Benveniste, director), the TFF of histamine caused coronary dilation up to 75 percent. From left to right, each vial indicates a minute of coronary flow: from the fifth, the increase induced by TFF is obvious; it reaches a maximum in the sixth and then decreases and goes back to normal in five minutes.

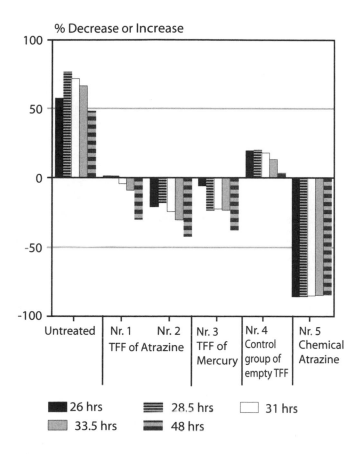

% Decrease or Increase

Legend		
■ 26 hrs	▤ 28.5 hrs	☐ 31 hrs
▨ 33.5 hrs	▬ 48 hrs	

Figure 6.3. The graph compares the photonic emissions of *Acetabularia* under the influence of poisons versus untreated control groups and groups treated with "empty" TFF (water): TFF of atrazine at two different concentrations (Nr. 1 and 2); TFF of mercury (Nr. 3); chemical atrazine (Nr. 5). Measurements taken between 26 and 48 hours show a clear drop in vitality of groups 1, 2, and 3, compared with the untreated controls and those exposed to empty TFF (Nr. 4).

in Paris at the laboratory directed by Jacques Benveniste. We used the "Langendorff Model" on an isolated, living heart by perfusing it with a Krebs-Henselheit solution. The infusion of 5cc of TFF of histamine in physiological solution was able to increase the coronary flow up to 75 percent compared to the control saline solution tested in parallel on another isolated heart, and compared to the 30 to 40 percent vasodilation obtained by Benveniste with homeopathic histamine (30 CH). The

Figure 6.4. Direct transfer to cultivations of wheat through metallic plates. The coleoptiles on the right (TFF of 2,4-D) grew more compared with the controls, because of the transferred auxinic effect.

TFF of atrazine acts as a poison against unicellular algal cells that are sensible to the toxic action of this substance (fig. 6.3).[13] We measured this using photomultipliers at the Biophysics Institute of Kaiserslautern, Germany, directed by Fritz Albert Popp.

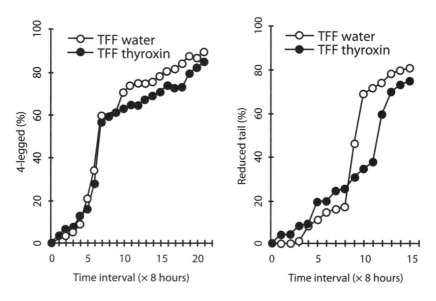

Figure 6.5. Two phases of tadpole metamorphosis—growing four legs (left-hand graph) and tail reduction (right-hand graph)—treated with TFF of thyroxin (shown by solid black dots) and empty TFF (water) as a control (white dots).

We also documented the effects of TFF of different types of substances with various properties such as herbicidal action, antigermination, or growth enhancement on plants cultivated in vitro (fig. 6.4).[14] The action of the TFF of thyroxin on the metamorphosis of the tadpole was reproduced and published with Christian Endler and Waltraud Pongratz (Graz, Austria) (fig. 6.5)[15] as well as that of the TFF of an anti-inflammatory in postsurgical pain in horses[16] and that of TFF of antibiotics in thousands of milk-producing cows suffering from mastitis.[17]

By pure serendipity we also made an unexpected discovery, in line with that of Sir Chandra Bose: we observed that metals, certain alloys in particular, keep the memory of information just like water. Information transmitted with TFF to metallic plates was retained by them and was not canceled by chemical agents. This indicates that, as the Italian physicist Luigi Borello wrote, all matter has memory of everything, not only water or metals, but stones and everything that exists.[18]

In TFF the chemical structure of the molecules to be transferred is important: the best results are obtained with the "aromatic" substances that have at least a benzoic ring. This is a ring of six carbon atoms, around which an electronic cloud is formed; the electrons of six atoms behave as if they belong to an ensemble. It seems that this favors the emission of signals from the substance, signals that are more intense and easier to transfer. The information is related not only to the type and the number of atoms of the molecule, but also their spatial configuration.

We have also discovered that the TFF can be affected by the disturbing action of electromagnetic fields; therefore it needs the protection of a Faraday cage, a metallic net connected to the floor, which isolates whatever is within it from external electrical fields. It also needs a mumetal type of screen.* On the other hand, it should not be isolated from the earth's magnetic field or from the cosmic waves that seem to be playing an important role, just like light and heat.

Some experiments were also done on water treated with TFF,

---

*Mumetal is a metallic alloy (made from 77 percent nickel, 15 percent iron, 5 percent copper, and 4 percent molybdenum) characterized by a high magnetic permeability and used to screen magnetic waves.

demonstrating that its chemical-physical nature changed after the treatment, confirming the reorganization of the water by the information transferred. In other words, after the TFF the water is not the same. The heat and conductivity of the TFF-treated water compared with controls were measured, showing an increase of many microwatts of heat and significant increases (in micro-Siemens) in conductivity.

All the physical and biological experiments with TFF employ similar controls of untreated water and a double check called *empty TFF*. It has to

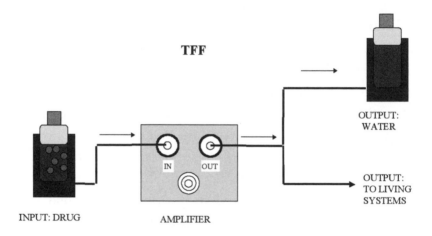

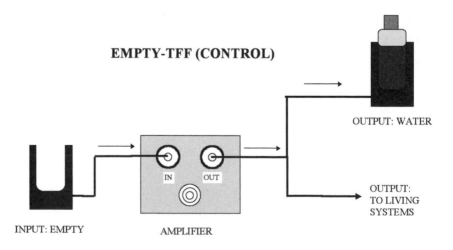

Figure 6.6. Difference in the preparation between a TFF of a medicine and an empty TFF, to be used as a control.

do with a transfer identical to the TFF, but without anything in the input of the unit, that is, a transfer of *nothing*. This double check is important to exclude interferences from the apparatus or the frequency emitted from the generator (fig. 6.6). Obtaining results with the TFF and not with the empty TFF means that only the waves emitted from the medicines are acting. The empty TFF is for us what in homeopathy dynamized water is: water that undergoes successions without having anything dissolved in it.

Placebo and Collateral Effects

In our experiments with TFF, the possibility of a placebo effect is excluded when a preparation is given to patients who, for example, are unaware of the nature of the drug or are unaware that they are not receiving heroin, and in blind trials carried out in animal and plant experiments. TFF preparations are devoid of toxicity since there are no molecules, and unpleasant side effects have almost never been reported, perhaps because of the hypothetical regulatory mechanism mentioned earlier. If you take an antihistamine tablet, you can get the undesired side effect of drowsiness, but with a TFF of an antihistamine, this side effect is manifested only in anxious patients suffering from insomnia.

The basic code derived from the informed field of the medicine also includes the information that can induce side effects, but the organism makes a selection of frequencies in which it discriminates between what is useful and what is not. When treated with a TFF, a body takes into account only the frequencies with which it resonates. Our hypothesis is that the selection is made by the organism's system of intrinsic regulation (SIR), which is able to discriminate among the range of frequencies that come to it (not only from medicines) and resonate with the useful ones. This system (which we will discuss in the next chapter) would be responsible for defending the organism against continuous electromagnetic "attacks" coming from the outside. We contend that, parallel to the homeostatic regulation at a molecular level, there must be a control center for frequencies that, self-regulating, select the useful information at that time.

The only case we observed of subversion of the SIR was in a schizophrenic patient: the first time we transferred the frequencies of an antipsychotic (haloperidol), the hallucinations subsided, but she had all the side effects of the medicine (from the stiffness like Parkinson's to depression), whose intensity prevailed along with the therapeutically beneficial effects for the entire first month. Then the side effects went away forever, and only the therapeutic ones remained. The patient continued the therapy with TFF for many years afterward. The occurrence of the side effects could have been due to the awakening of the memory of the chemical medicine in the body, induced by the frequencies of the TFF. Or maybe the SIR of a psychotic person takes longer to recognize useful frequencies?

Of particular note are the reactions we observed in youngsters using hallucinogenic amphetamines (so-called ecstasy), on whom we tested the TFF of that substance, with the aim of curing them of the negative effects of the drug taken up to that moment. We prepared a wave inversion TFF of ecstasy, thinking that, according to Morell's principle (that reversed waves induce inverted biological effects), it would cancel out the effects of the drug.* Contrary to what was expected, the TFF didn't invert the effect but reproduced it instead, so much so that the youngsters wanted more. Then a TFF of amphetamine was made without inverting the wave, which produced similar effects to that of the original drug. The same results (with normal TFF effects or in inversion) were absent in people who never used that type of drug. Maybe in the case of drugs or other substances that alter perception and thought, the TFF acts only when there is memory of previous intakes. A few minutes after the TFF intake, the youngsters felt the symptoms of the ecstasy "coming back." Or was it the TFF evoking the memory? Or could the system of intrinsic regulation have altered the drug itself?

---

*The principle of the wave inversion is not new: the headphones of helicopter pilots receive the noise of the engine in reverse phase to cancel that of the engine itself. A project is studying what are known as "islands of silence," city neighborhoods where disturbing noises would be defeated by sensors that, after having recorded them from the environment, would return them inverted in phase.

Some experiments have produced opposite effects to those expected. This happens for many reasons: one is the concentration of the medicine used. This indicates that the same medicine, at different dilutions, emits different signals. It is as if, instead of a soprano singing the solo of the Queen of the Night in Mozart's *Magic Flute,* there was a chorus of baritones accompanied by a fanfare: the notes would be the same, but not the result. In a similar way, the gurgle of a medicine can become a raucous baritone.

Another cause is the length of the cables used in the transfers: if they are too long, they can reverse the signal. In support of this theory are the works of the German biologist Michael Galle, who obtained opposite results by growing wheat with TFFs of herbicide prepared normally and in inverted phase.[19]

It also seems that lunar phases affect the transfers, which is logical, considering that we are talking about liquids. Even the daily transits of the moon are detected by water (with variations of some picoamperes), depending on whether or not the moon is present in the celestial hemisphere at the loading moment (fig. 6.7).

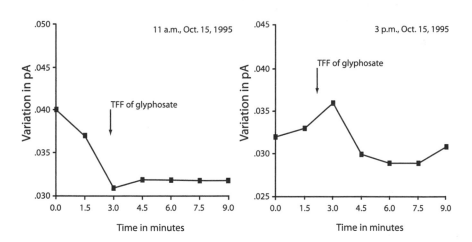

Figure 6.7. Variations due to lunar transits. The analysis of the variation (expressed in picoamperes, pA) of the information supplied by the water, after the irradiation with TFF of glyphosate, shows different results according to whether the moon is present (11 a.m.) or absent (3 p.m.).

Applications of the TFF

The TFF can have important practical uses in at least three industrial sectors: food, pharmaceuticals, and agriculture. By irradiating foods subject to easy deterioration with frequencies of antimold substances, we could prolong their preservation without altering their authenticity, a true ecological sterilization. German colleagues have successfully applied the actions of antiparasitic systems on vegetables by feeding them water with inversions of the frequencies emitted by the parasites.

Many drugs used in the pharmaceutical industry for human and veterinary pathology could be substituted with frequencies transmitted via cable or diffused in the environment or transported as solids or liquids: antibiotics, anti-inflammatories and analgesics, antihistamines, benzodiazepine, hormones (particularly estrogens, progestins, and thyroxin), and other medicines. The use of TFF of dopamine in the treatment of Parkinson's disease and the use of TFF in the cessation of heroin use are particularly interesting applications. These benefits come without toxicity and with economical savings.

The pharmaceutical industry would benefit from the production of waters impregnated with therapeutic information as well as apparatuses that memorize the frequencies of medicines. By perfecting the transfers, we can conceive applications for an indirect way (irradiation of therapeutic frequencies in the environment), then the waves of the medicines could be electronically synthesized. An interesting area to experiment with would be TFF in chemotherapy, particularly to verify if in cancer cases the SIR can select the information to destroy the tumor cells without selecting those that produce the side effects. Following the principle stated by Morell, which attributes therapeutic quality to inverted wave phases of certain pathological frequencies, it would be interesting to experiment with patients after they have been operated on for cancer by supplying them with a TFF of inverted oscillations emitted from the cells of their own removed tumor.

In agriculture TFFs of herbicides and pesticides could be applied via environmentally friendly solutions soaked with these frequencies. We should be able to achieve effects on crops without producing toxic residues and environmental contamination. True organic farming.

7 ○─● FOR A SCIENCE OF THE INVISIBLE

Have you ever wondered why an entity is always as it is without losing its identity? Why does iron remain as iron and not change into copper or lead or wood, and why does gasoline remain so rather than turn to liquid nitrogen? Nature jealously guards the identity of any substance, and it remains itself for all time. So do the forms of a body's architecture: stones, crystals, objects, plants, and animals. With difficulty we would observe a pencil case transforming itself—let's say—into a glass, or into a bottle, or into something unknown. You wouldn't even see it melting like snow in the sun, for example, slowly losing its shape, becoming mass without specific identity.

It is an extraordinary thing that bodies retain identity and form, since they are made of parts (atoms and molecules) that could be uncombined and then recombined to form different shapes. Why doesn't it happen? If their matter comes from a common nursery of particles, why does an apple differ from a crystal? Who ordered them like that? What molds the shapes? If they were vibrating strings, who would play them? How is the rhythm established? How is it kept?

A minimum scrap of charges distinguishes the bricks of the universe; only the gradual addition of electrons causes their differentiation. We would only have to add an electron and a proton to potassium to transform it into calcium. But this does not happen. It can, but only in

91

certain rare conditions. A camellia will bloom, make flowers, and die, but it will stay the same. An acorn will become a bud and then an oak: it evolves without losing identity. It will never happen that a rose will become an oak or will transform into a block of pyrite or a kangaroo. Why? What keeps the matter in the confines of the shape? What allows the bodies to preserve themselves? There must be something that gives them their border, a mechanism of regulation.

This is also true for cells: some system monitors and controls cellular homeostasis, and it checks the biochemistry. Have you any idea just how complex the network of chemical reactions within a cell is? It is far more complex than that of a computer circuit. Is it possible that this complexity is lacking a director? Who could be the director? Not the DNA. Despite their complexity, nucleic acids can't have this function. There must be something to guide them, some kind of regulation circuits for every cell, for the organs, the systems, and indeed the whole body.

Fritz Albert Popp wrote that the models based solely on receptors couldn't solve the fundamental problems of regulation because "every regulator needs in turn a new regulator and so on, infinitely"[1] There needs to be something at the top of the pyramid acting as the director: the ultimate regulator, intrinsic to the system (not transcendent), which could even identify itself in the system, in its entirety and complexity. This is what makes a system of self-regulation plausible, an SIR (system of intrinsic regulation) for each body, cellular or otherwise, expressing the basic code related to the body over which it has control.

In order to understand the concept of a system of intrinsic regulation, let's proceed step-by-step, starting again from the elements. Imagine having an atom in your hand and looking at it attentively: its characteristics are given by the numbers of electrons and the distances of their orbits, which and how many particles form the nucleus, and so on. The balance of the electrons is essential, for if they were too close they would collapse on the nucleus, while being even a little bit too far away would cause them to shoot out into space; in both cases the atom would break up. The balance between the charges is only the mechanism, not the design. We have to discover what establishes the right distances between electrons. Matter is organized with mathematical precision, more the expression of a design than causality. Considering chance in this scenario is crazy: it would be like believing that—to cite a well-known Latin author—if millions of letters of the alphabet written on scraps of paper were launched into the air, they would fall back on the floor, assembled in such a way as to rewrite the entire *Annales* by Ennio.

Here we enter the domain of the maps of things. Chemical formulas are not enough to account for the anatomy and the physiology of bodies without the intervention of ordering principles and control mechanisms that are not molecular. For every entity there is a design template, a basic code, and a guideline for forming and maintaining the structure. The chemistry describes how and of what things are made, not how the ingredients are chosen, how they are mixed, in what measure and within what limits. There is still no answer to the ancient enigma of form and substance. We still do not know enough about what bodies are made of, about what exact criteria inform the choice of elements, and about how they are to be positioned.

We have seen that any construction is preceded by a design. Mother Nature does this with the codes, starting with pure matter, which is not yet determined and is therefore the mother of all things. Pure matter is the "immaculate conception." What Virgin Mary represents is the Matrix that has not yet been "contaminated" by becoming this or that. In addition to Virgin Mary, the Matrix is the Great Mother in every era and culture: Ishtar, Isis, Astarte, Cybele, Demeter, Ceres, Brigit, and all others. So the first question is how a single form arises from the Matrix. The second is what mechanisms allow the bodies to retain form and identity. Let's now find some answers.

We have seen that a thing emits information into its field. Our work with TFF, in which we have recorded and transmitted that information, has affirmed its presence in the field of each thing. That information is a necessary aspect of its own formation, that which "in-forms" it. It is that which acts as an architectonical schema from which the body derives its own structural references. There are some other functions. One is the homeostasis that guards and preserves what has been built. Once the construction of a house is finished, the schema ceases to be useful, whereas a natural entity requires its schema to be constantly present, otherwise it would disintegrate. If the electromagnetic component of the field is generated from the matter, it is also true that the matter is produced and maintained by the informational component of the field, by the continuous homeostatic process between fields and masses that inform one another. "It is the energy fields that generate the matter and not the opposite. They proceed and organize the formation of the physical body," writes American colleague Richard Gerber.[2]

How do we imagine the basic codes? Seeing that in the universe everything vibrates, we can think about them as oscillations or rhythmical pulsations specific to a particular substance: hydrogen pulsates at a rhythm, oxygen at another, water at yet another, and so on. From these pulsating rhythms, changes in the basic rhythm of matter, foamy waves emerge from a peaceful sea. The physical bodies are nothing but expressions of frequencies, numerical sequences arising from the alternation of impulses and pauses. Intermittent signals, waves, and frequencies guide the architecture of the bodies in the direction foreseen by the basic codes to achieve a balance of charges and forms; matter is formed by the continued flux of information, the music of creation. It is through the music of the planetary spheres, such as the electronic clouds, that Mother Nature carves matter, making it perceivable. We are back to Orpheus.

The code generates the combined matter and then preserves its identity and form. The information acts as a unified force field, constant in its rhythms of pulsation, and as long as these rhythms are not altered, the matter remains itself and returns that information to the field in a process of continued self-regulation and indeed self-expression. Think about carbon, imagine it as dioxide embedded in a sugar molecule or in the paper of this page or in the fossil imprint in a rock. In any case, its code sends information to the molecules to place themselves in such a fashion as to constitute carbon: the number of atoms, protons, neutrons, and electrons it must possess, how they should be organized in the nucleus, at what distances they should orbit, how to rotate, with what *spin*. It communicates all of the chemical and physical characteristics that make sure that the substance is carbon and nothing else. The physical structure of the carbon is maintained by the signals that its code exchanges with the matter: the information travels from the code to the mass and back to the code again, like self-reflecting mirrors. Without basic codes, any body would disintegrate because it would lack its own references.

It is a wonder that table salt does not separate, disintegrate, or transform. It is all because of that program that governs from deep inside its atoms, in a space apparently empty, but instead one that is in truth the breeding ground of everything. Sodium chloride does not spontaneously disappear into sodium or chloride since the code of sodium chloride takes precedence over others. The salt remains salt, with all its properties. If instead, as Giordano Bruno wrote, we could modify the code, we could mutate a thing into another, as seemed possible for the ancient alchemists. They did it with metals. These are not fairy tales, but relate to recent events, from the last century.

Back to Marconi

I met Pier Luigi Ighina. I am the last man on Earth who interviewed him, a month before he died in early 2004; he was 94. For more than a decade he had worked with Guglielmo Marconi as a technician and his apprentice. He was not just a mine of knowledge and an author of brilliant discoveries, he was above all a great humble master of science and life.

At the age of sixteen he discovered what he called *magnetic atoms*. He could see them thanks to the atomic microscope he invented, which magnifies by 1.2 billion times. Based on his observations of atoms with his atomic microscope, he defined four fundamental laws pertaining to all atoms:

1. When light atoms excite the observed atoms, the observed atoms absorb some of their motion.

2. The observed atoms absorb part of the motion of the light atoms to speed up theirs.

3. In order to excite an atom, it needs to be in contact with an atom of higher motion; the atom with the highest motion will attract the one with the lowest motion.

4. The higher the atom's motion, the more luminous it will be, and vice versa.[3]

Ighina classified matter according to the different pulsations and absorption rates of the atoms. Although he found that magnetic atoms are in all matter, he described them as having special properties not shared by all atoms. A magnetic atom is smaller and faster; in fact it is in perpetual motion. Its pulsation transfers movement to other atoms. In order to isolate it he created a kind of "wall" composed of different layers of atoms, setting the ones with maximum absorption (95 percent) at the inside close to the observed atom, and then one after the other, different types of atoms with diminishing absorption rates (85 percent, then 75 percent, and so on until 1 percent).

With this idea he was able to create "canals" that would take the motion away from the observed atom until it was almost still. They show up in the photograph he took of a magnetic atom in his laboratory in 1940 (fig. 7.1). The observed atom radiates pulsating magnetic energy into

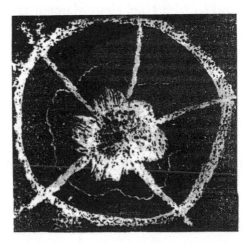

Figure 7.1. The magnetic atom (from P. L. Ighina)

the surrounding space, visible—in the picture—as a thin luminous circle around the central atom.

Putting a magnetic atom in contact with other different atoms, Ighina observed the following: when the magnetic atom is isolated it develops its maximum motion until it meets another atom of its same motion sensitivity (pulsation); the contacted atom starts to move and absorb pulsations from the magnetic atom until it has reached its maximum motion; at this point the two atoms split apart.[4] But, since its pulsation is everlasting, it quickly recovers, and the whole process begins again. In this way he discovered that the magnetic atom is the one that gives motion to all the others.

According to Ighina, the shapes of different entities derive from the alterations of the vibrations of their atoms. He thought that the influence of the magnetic atom on the vibrations of other atoms indicated that it was responsible for all the variations of all atoms. So he developed an apparatus that enabled him to regulate the magnetic atomic vibrations. Through various experiments with it he discovered that he could change one formation of matter to another. Here is the description in his own words:

One day, once the regulating apparatus was tuned in to the vibration of a given matter, I left it in the same position until the next day. In the meantime the apparatus had changed its own vibration and that of the matter was in turn increased. I noticed that this matter seemed now to have a different structure, more similar to that corresponding to the new vibration. Other tests made me understand that by varying the vibration of one kind of matter it could be transformed into another.

With the same apparatus, one day I discovered the exact vibration of the atoms of an apple tree, and of a peach tree. I synchronized the apparatus with the vibration of the peach tree, then started to increase the vibration little by little until it reached that of the apple tree. The increase took eight hours, after which I kept the vibration of the peach tree at the vibration of the apple tree for sixteen days.

Little by little I saw the peach tree transform and become an apple tree. With the same system, a Mayflower peach, which is a small variety, could be transformed into a larger variety of peach.

I started to do experiments with the same system on animals and I thus managed to transform the tail of a mouse into that of a cat. The duration of the transformation of the tail lasted four days, then it went back to the original state but it detached itself and the mouse died. The atoms of the tail didn't take the alteration for long.

Even more interesting was the development of a bone of rabbit affected by osteomyelitis. The two ends of healthy bone near to the infected area had different vibrations. I tuned in through the apparatus with the vibrations of the healthy ends of the bone and I started to alternate to the maximum vibration, and I obtained the phenomenon of the reproduction of the atom. Soon, the two ends of the healthy bone came close until they met; this was accomplished by providing the continuity of the vibrations that had been inter-rupted by the infection. The bone went back to normal, as well as the vibration, and the fever disappeared.[5]

In the last interview of his life, Ighina was fully convinced that matter was rhythm that pulsates in different ways (a thought that is not so dif-ferent from the superstrings theories). To transform a peach tree into an apple tree it is sufficient to record the rhythm (the basic code) of the apple tree and irradiate a peach tree with those frequencies. By knowing the code of things and working with it, it seems possible to change matter.

How did Ighina record the code of a tree? Here's what he said in the interview: "Take an aluminum pipe, twist it to get seven spirals that you wrap around the tree branch and with an electric cable take it inside the laboratory." Once he got the frequency, he could reproduce it so that another plant species could listen to it until it transformed itself into the initially sampled species. In other words, Ighina recorded the rhythm of the apple tree by capturing it from the field of the tree, con-firming that the information radiates in the surrounding space. He was performing a TFF from plant to plant. His experiments also confirm that by possessing the code of a substance, we can do what we want with it. Here physics is combined with magic and alchemy, which perhaps are three different aspects of a single thing.

As far as alchemical transformations are concerned, other testimonies come from the French doctor C. Louis Kervran, a member of the Academy of the Sciences of New York, who in the late 1950s had documented a case of spontaneous transformation of elements. As mentioned before, adding a simple electron to potassium will transform it into calcium; this transformation naturally takes place in some living things. Chickens subjected to a diet completely devoid of calcium produce eggs without shells, while a diet without calcium but rich in potassium immediately brings the shells back. The alchemy of the organism transforms the potassium into calcium.[6] The SIR of the chickens *knows* that there is an absolute need for calcium and tells the potassium to transform itself into calcium.

But let's go back to Ighina, who photographed that particular atom that sets the rhythm as a pacemaker. Its pulsation transfers its characteristics. What he calls the magnetic atom could be an expression of the basic code that informs the field ("it pulsates and radiates energy in the surrounding space"), whose vibration the atoms record so they can express substance and quality together. It builds the shape. When Ighina observed the perturbation of space, he expressed it in a way sounding like the basic code: "I noticed that each type of matter has its own magnetic field composed of magnetic atoms and atoms of the matter itself."[7]

The basic code would be the inexhaustible source of the information that organizes matter, powerful enough that, transmitted to a second substance, it takes the characteristics of the first. In the TFF, information is transmitted just like in the experiments of the follower of Marconi, such as transferring to a mouse the information of a cat, so that the code of the mouse is modified into that of the cat—"biological TFF"! When a code is transferred, matter is transformed.

The same happens with abrupt alterations of magnetic fields and electrostatic variables recorded near volcanoes in geologically active areas, especially before earthquakes. Earthquakes are preceded by intense fluctuations of the geomagnetic field, orthogonal microwaves near the terrestrial surface that, according to Giovanna De Liso, can be seen as indicators of their occurrence. She has experimentally investigated variations in the electric conductibility of rocks and their magnetic permeability, which form between them an electrostatic field, like in a condenser, in connection with the so-called Shroud effect. The Shroud effect refers to whether images of objects lying between the two folds of a linen cloth of characteristics similar to those of the Turin Shroud, smeared with different solutions (aloe, myrrh, blood in a water/oil solution) and placed between two horizontal gneiss rock layers in an underground cellar partly excavated in ferromagnetic rocks, could form naturally before or during an earthquake. Among the leading experts on the historic cloth kept in Turin, whose image seems to refer to Jesus, De Liso managed to imprint images of different objects by using the technique just described (may we remind you that the Gospels and pagan literature speak of intense earthquakes at the moment of Jesus' death on the cross and in the following days).[8]

This is almost a geological TFF. In our transfers, the information is transmitted in a similar way, with variations of electrostatic fields, just like in a radio. And so we are back to Marconi. TFF, radio, and the Shroud effect all seem to testify that, through field variations, nature can transfer information from one system to another. These transfers can also be translated into sounds and images.

Communication between Things

Something similar to the "transformations" made by Ighina was realized by the Chinese scientist Chang Kanzhen, together with Gregory Kazmin, director of the Institute of Research for Agriculture in the Far East. Kanzhen built a transmitter of biomicrowaves that transferred information from one living being to another, plant or animal. By irradiating seeds of corn with informed waves from wheat, the corn plant produced grain similar to both wheat and corn. From chicken eggs irradiated with the information of a duck came hybrid chicks.[9] It seems that a growing organism manifests aspects of the form corresponding to the irradiated information. The experiences of Ighina and Kanzhen support the hypothesis of the basic code, responsible for the forms transmitted to other bodies, almost a TFF in a cellular environment.

The informational essence of the basic codes is communicated by sequences of impulses and pauses that mark the vibrations of the bodies, which teach the matter in which rhythms to oscillate. Different things exist because they vibrate at different frequencies. The codes set the rhythms by which things are differentiated. The space surrounding the body is never still, and the information in the field is the same that vibrates within the body. It is able to interfere with the environment to inform it. The information is spread from a body in an incessant and continued way, without ever being exhausted or consumed, as long as its basic code exists.

Thus bodies are more extended than how we usually perceive them: they have an invisible part made up by the informed field arising from the basic code. They are pulsating dynamic structure—*motu proprio*—which the senses don't perceive. The information that organizes molecules propagates through the perturbation of the field to the structures of the body in order to communicate with the environment. It is the emanation we record from the medicines and that Ighina captured from the branches of the tree. This is how bodies "communicate." Let's try to find out how.

Theoretically, the informed fields of two entities can interact in a simple way or in a transgressing way. In a *simple interaction* between two bodies in contact, information about each one's identity is exchanged through their fields, and both substances gain experience of each other (fig. 7.2).

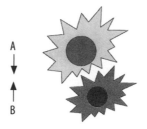

Figure 7.2. The simple interaction model between the informed fields of the bodies *A* and *B*. The exchange of information is reciprocal, and each field continues to be informed of the other for a time.

Two people meet—"Good morning, I am Tom. Nice to meet you, I am Dick"—and they define themselves by sharing their codes. An object relates to another about its qualities (shape, matter, color, consistency, structure, property, and so on), and vice versa in the same way. Through the basic codes, bodies are always informed of what happens in their surroundings.

In water, the field variations are better fixed and last longer. Schimmel's experiment, mentioned earlier (in chapter 5) demonstrates a simple interaction in which the homeopathic phial transmits information to the water in which it is immersed and the water becomes a copy of it. Water could also transmit its code to the phial, except that water acts as a universal receiver: it can receive imprints of everything without giving anything of itself. Also, even if there were a transfer of water to the remedy, it wouldn't be possible to verify it.

Let's look at another situation. Again, let's consider two people introducing themselves: one shouts his name in the other's ear, and she is stunned and overwhelmed. It is not an exchange anymore, but a monologue, a one-way communication. It is a *transgressing interaction* when the field of a substance is powerful enough to make the emission of its signals more intense; once they become sufficiently penetrating, they transgress the field of the other substance (fig. 7.3).

The transgressed field changes its rhythm and enters into a new matter state by becoming similar to the other substance. The level of resemblance depends on the information and the conditions of transfer. The stronger one transmits to the other until they oscillate together.

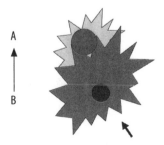

Figure 7.3. Transgressing interaction model. The activated field of substance B strongly imprints itself on the field of substance A.

The TFF we have performed on liquids is a transgressing interaction because the generator activates the information from the medicine, which makes the water vibrate to its frequency; in this way the medicine is *copied* into the liquid. Similarly, in the preparation of a homeopathic remedy, the succussion that transmits the oscillation of the solute to the water molecules creates a transgressive interaction.

The same interactions—both simple and transgressing—that govern exchanges between things are found between things and organisms. Simple interactions are demonstrated by medicine tests (such as kinesiology, electro-acupuncture, or auricular medicine) in which the organism "feels" the field of the medicine by reacting immediately. Communication goes the other way as well: a person can inform an object. If an object belonged to Tom before being used by Harry, the field of the object will retain morphological and emotional information from Tom as well as Harry. Those who are very sensitive or psychic and come into contact with an object are able to read life events or emotions of the people whose field has become imprinted in the object. This is termed psychometry. This is not parapsychology, but the physics of interacting fields. If the transfer is transgressing, memories of particularly intense emotions or of other dramatic facts can imprint into an object for centuries.

A Theory of Communication

"Our universe it is not all that makes up the cosmos. The greater reality is the Mother of our universe and maybe a large number of others, some prior to ours, others coherent with it," writes Laszlo.[10] He is convinced that even the communication between universes is made via an incalculable amount of information, perhaps through black holes, where spatial-temporal laws are not respected anymore.

Let us remain in our universe, where information is exchanged through signals. In the light of what we said before, the exchanges should take place in two ways: molecular action and field interaction. In the first, molecules with "strong interactions" at a short range transmit the messages. These are the chemical reactions that form stable ties with the vicinity. The kinetics of field interactions, whether simple or transgressing, are long-range "weak interactions"; they are able to act at a distance without forming chemical bonds. Two substances react chemically in close proximity; it is also possible for the weak signals of the molecules to produce effects across longer distances, with field actions. For example, we have seen that a medicine can be administered in both ways: in molecular kinetics, the cell receptors are stimulated according to the model of "key in the lock," where the drug acts on the receptor molecule as a key, which is able to open it; in the field of kinetics, however, the receptors are activated by the signals of the medicine, like a remote control of the lock.

The action of the field is masked by that of the molecules, and in some cases occurs beforehand, from the first contact of a medicine with the patient taking it (because the signals don't meet any obstacles and they can't be modified). Lehninger* observed that identical biochemical reactions are faster in vivo than in vitro. This may be due to the fact that, in the organism, the weak action of the field is amplified in relation to the number of cells, becoming faster and more efficient in complex systems. This is what we have observed in the effects of TFF on multicellular organisms compared with unicellular. This is in accordance with the Kaiserslautern postulate, which states that the more complex the system, the more complete the decoding of the signals. According to complexity theory, a system designed as an aggregate of parts interacting with each other can have different behaviors than the individual parts themselves. In other words, an organism may react differently from some of its cells within the whole.

Because of the kinetics of fields, a drug can act even in the absence of molecules. The opposite, namely action arising from a molecule disconnected from a field, is not possible, however; having lost its own structural references, the matter would break up. While there may be fields devoid of molecules, it is not possible to have a body devoid of a field.

---

*Albert Lester Lehninger (1917–1986) was an American biochemist, widely regarded as a pioneer in the field of bioenergetics. He made fundamental contributions to the current understanding of metabolism at a molecular level.

Psychometrics suggests that bodies can "memorize" structural and emotional information from beings that they come into contact with. This topic will be dealt with further on in the chapter on emotional fields (chapter 11); for now we will talk about the contact itself. The word *contact* is derived from *con,* "together with," and *tangere,* "to touch." When two human hands touch, not only are the touch receptors stimulated, but there is also a meeting between fields. The two phenomena overlap, making it difficult to notice the field interaction. When a hug or a handshake is exchanged, the essence of each person—his or her entire code—is communicated to the other, making it possible for them to *read inside* each other. But only someone who is sensitive to field changes can use that awareness to perceive illness or the particular moods of others. The aforementioned words of Jesus, who knew that virtue had gone out of him when his robe was touched, can be interpreted as a reference to the contact between fields.

The physiology of the senses should be reviewed in light of the physics of the informed fields. For example, the sense of smell seems stimulated by volatile molecules, but this does not explain some sensory situations. How do you explain that a sniffer dog can find minuscule quantities of a drug wrapped in many layers of clothes inside a case, only on the basis of the molecular model? Even admitting that some molecules might spontaneously detach themselves from the drug, how could they escape the plastic bag, go through layers of clothes, and pass through—God only knows how—the side of the case to fly freely in the air and hit the receptors of the mucous membrane of the dog? It is against the principle of physics. It is like a *tunnel effect,* a quantum phenomenon where a particle is able to go through an energy barrier even if its energy is less than is needed to do so from the classical physics point of view. What energy could make the molecules overcome so many barriers? Generated from what?

It seems far more plausible that the field of the drug became resonant with the layers of clothes, the leather of the case, the air, and the sensory receptors or the field of the dog itself. Where a molecule is unable to go, a signal can penetrate. As we have seen, the signals of medicines pass through glass. Those of dopamine reach even the midbrain of the

Parkinson's sufferer, whereas dopamine molecules are blocked by the blood-brain barrier. If the senses are stimulated not only by molecules but also by signals from fields, a dog trained to a particular resonance of a particular drug possesses receptors able to resonate with the basic code of that substance.

Thanks to the long-range field reactions, similar molecules can *feel* at great distances and vibrate at the same frequencies, even if other molecules are between them. Oscillating the molecules at the same frequency for the duration of the biochemical reaction is like dialing a number with a phone and maintaining that connection for the duration of the call. Nothing can distract those two that are calling each other. The theory of superradiance, describing the effects generated by particles of coherent matter oscillating with their electromagnetic fields, states that a messenger molecule introduced into a cell goes directly toward the target, as in a transmitter-receiver connection. On the other hand, traditional biochemistry suggests that, before reaching the target, molecules would sustain incalculable and unnecessary collisions with others. The traditional model expresses mechanistic thinking, which is complicated beyond belief. Nature does not waste time and energy in attempts if it can achieve safe results immediately and with minimum effort.

Psychometrics and the compatibility tests of medicines confirm exchanges with contact or minimum proximity. Long-range interactions can be seen in processes like dowsing, in which messages from underground are received; this takes place through resonance with the fields of water, oil, or other mineral layers. The literature on the phenomenon of *nonlocality* documents occurrences of simultaneous influence in spatial conditions that would make any communication impossible. They are explained by admitting the existence of a background continuous connection of their fields.

At this point, we must take another break to observe the images that, as in the kaleidoscopes of our childhood, can be imprinted on water and on matter in general: the embellishments, arabesques, and even crystals that music can imprint into things. We return to talk about water and what it has to do with lights and sounds.

8 ⊶ LIGHT AND MUSIC ON WATER

Music enchants matter. It literally *creates through singing*. Let us now explore how sounds are able to inform matter, transforming it to produce shapes, images, music, and light. The basic codes that direct molecules and govern bodies are music, but they are infrasound, which we cannot hear. Though it may seem to be a shame that we are unable to listen to the symphonies of things, this music may not be as harmonious as that of our musicians. In fact, it is precisely because we cannot hear these sounds that we can function. Imagine an existence where we heard every single noise, from the vibration of our body molecules to waves of all kinds and the noise of the stars. Dazed day and night, we would not understand anything anymore; life would not be possible. That which is audible to the human ear is the necessary threshold for our survival.

If we dare to go beyond the limits of the filters of Mother Nature, if we go beyond the boundaries of the audible frequencies, we finally understand that the entire universe is music. And what is music? Rhythm. That reminds us that Pier Luigi Ighina told us that matter is rhythm. And rhythm itself? Sufi master Abd alRazzaq alQashani wrote in the fourteenth century, "Let's take a metaphor. The land that is hit by the sound wave is itself an undulating movement. The wave is the meter, the rhythm is the combination of the tones on this wave . . . The tones are distributed on the measure regularly or irregularly; they can occur in rapid succession, or on the contrary leave vast empty intervals. Sometimes they pile up; sometimes they are distant from one another. These games of the tones on the sound wave, this model of the substance of the wave . . . this is what we call life."[1]

Now let's consider what Pythagoras (who rejoiced in numbers and sequences) did with a simple string. He fixed it at both ends, plucked it, and listened to the sound. Then, he divided the string in half and plucked it; he heard the same note as before, but an octave higher. (It is called an octave because it is the eighth note in a seven-note scale.) He also discovered that the most pleasing harmony occurs when the ratio of the two lengths of string is 3:2 or 2:3. (We call this note a fifth because it is the fifth note of the modern Western scale.) This is what happens when a string lands in the hands of someone like Pythagoras—the musical scale was born.

Observing the Waves on Water

Rhythm is a fragmentation; it is an alternation of tones and pauses, *solids and empties*. The pauses define the rhythm, the help empties to build the shapes. Mozart—an adept of alchemy and a Freemason—wrote that real music is *between* the notes. Once again we glimpse the other side of things, which expresses what is in *between* the solids, as in between the notes. It is into this world of fluxes and waves that we will go to investigate the nature of things further. The science that deals with waveforms is called *cymatics* (from the Greek κύμα, "wave"). Vitruvius, the architect of imperial Rome, was the first to investigate the analogy between the propagation of sound and the movement of the waves on a pond; he inaugurated the study of the relationship between sounds and water. At the end of the eighteenth century the musician and physicist Ernst Chladni discovered the relationship between vibration and form. His experiment was simple (the simplification favors the truth): first he placed some very fine sand on a metallic plate; then he played the violin perpendicularly to the circumference of the plate.[2] While he was playing and producing sound waves, the sand, as if guided by an invisible force, assumed symmetrical shapes in response to the sounds of the violin. Thus were born the first geometrical shapes described by Chladni that bear his name (fig. 8.1).

Sound waves can restructure masses to produce coherent and ordered images. This discovery must have seemed odd at the end of the 1700s, but let us not forget that it was during the Enlightenment that research on sounds and colors started.[3] Sounds move matter by orienting it to produce new ordered shapes, just like in the chemical reactions discovered by the chemist Raphael Eduard Liesegang: if some drops of silver nitrate are poured on a glass plate spread with gelatin and potassium bichromate, the silver chromate collapses in concentric rings toward the outside, now called Liesegang rings.

Chladni's studies were resumed in 1815 by the American mathematician Nathaniel Bowditch, who discovered that the waveforms were

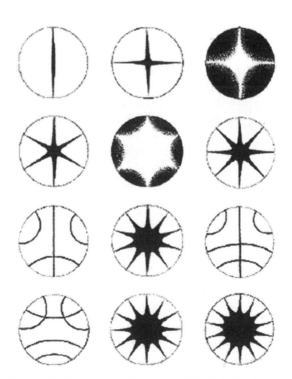

Figure 8.1. Examples of images Chladni obtained on metallic plate (from A. Forgione)

created by the intersection of two sinusoidal waves generated by two different sources perpendicular to each other. The so-called Bowditch curves can be observed today through oscilloscopes. Even the French mathematician Jules Antoine Lissajous carried on with this type of research in 1857; he observed that two different categories of images could be obtained, corresponding to the two waves being in phase or not. If they were not in phase (because of different frequencies) more or less harmonious image networks would appear. If they were in phase, the result was a circle. In 1827 the physiologist Charles Wheatstone built a *kaleidophone* with little pointed metal rods with polished edges, which reflect the light from a source such as a candle, generating different shapes according to the speed of the vibration.[4] All these experiments were useful to understand the close link between light vibrations and sound on one side and the organization of combined matter on the other.

The leading researcher in the field of cymatics was the Swiss physicist Hans Jenny; it was he who used sound to produce formations on the essence of turpentine (fig. 8.2). Jenny writes that all cellular life is

Figure 8.2. Essence of turpentine excited by certain sounds (from F. A. Popp)

governed by rhythms, periods, cycles, frequencies, and sequences; it is music, it is the "creative style of nature."[5]

It is not unreasonable to consider matter the result of pulsating rhythms if, as Jenny writes, "the harmonic systems are reproduced through oscillations, that is, harmonic intermittences, and the forms are the result of rhythms and intermittences."[6] Pulsating fields guide matter in a rhythmical way so that it takes certain shapes. So many analogies can be seen! The configuration taken by the chromosomes at the moment when a cell divides to multiply itself is the same as a scheme of the electric field of an electromagnetic wave aroused in a resonator with coaxial cavity! J. Hartmann discovered that the electromagnetic fields of certain geopathogen areas (variations of the geomagnetic field, potentially harmful to health) can destructure images of crystallizations.[7] Therefore, we can deduce that some electromagnetic waves, just as some acoustic ones, are able to record ordered structures in matter, each of which is a result of the information and in turn the message.

Jenny, using crystal oscillators and an apparatus of his own invention called a tonoscope, managed to generate modulated sequential sounds and to observe their effects on mediums of different types (fluids and thin dusts). He also studied the effects of the vibrations of different languages. If the sounds emitted are from ancient languages, such as Sanskrit for example, the shapes obtained are reflected in ancient traditions, as are the meanings ascribed to them. For example, when the seed dust of the *Lycopodium* fern is spread on a crystal glass and then exposed to the sound of *OM*, the most famous mantra used by practitioners of the Eastern tradition and meditation, the prolonged sound generates, in the dust, the image of a circle with a central point. This image has always represented the sun, the creative force of the universe, which is the meaning of the mantra itself.[8]

Jenny also collected a lot of material relating to the temporary images produced in drops of water. Under the microscope, tests were filmed documenting how a single water drop reacts to infrasound (Fig. 8.3). In both the dust of the fern and the surface of water, Jenny was able to obtain images that look three-dimensional; their complexity and beauty corresponded to the type of frequencies used. Starting with low sounds and gradually increasing the frequency, he obtained shapes more complex and ordered, similar to those of Pythagorean geometry, which Pythagoras defined as "solidified music."

Besides the ordered geometric shapes, schemes such as that of living cells of more complex organisms could also be reproduced. For example, a certain sound frequency could get the water drop to look like a maple tree leaf, while another would transform it into a coin.[9] It was these similarities between forms in nature and those induced by sounds that suggested to Jenny that sounds (and vibrations in general) play a role in the organization of the structural schemes of living beings.[10] The shapes lasted only for the time that the sound was active; as soon as it ended, the images would disappear. This agrees with our conclusion that sounds create and organize matter according to the information they carry: the shape remains as long as the code remains, and the dissolution of the basic code disintegrates the structure.

Figure 8.3. Reactions of a water drop to different sounds. At low frequencies, the molecules start to curl, vibrating in simple circles or concentric ones; if the infrasound frequency is increased, the complexity of shapes increases (from A. Forgione).

Everything Produces Sound

Sound creates, sound destroys: it was the sound of trumpets that destroyed Jericho's walls. Life and death are the umpteenth alternation of solid-empty that characterizes the world stage. "Everything in the world is inconstant, because everything changes. Only one thing is constant and that is change itself: the Tao"; this is how Lao Tzu started his *Tao Te Ching,* 250,000 years ago. At any level, the universe is rhythm, pulsation, and its codes are sequences of changes.

What Chinese Taoists call Tao, Indians call Shiva, the god who creates and destroys the world by dancing. Shiva is the cosmic dance, change, rhythm, the process of creation and also destruction. Alexandra David-Neel writes, "All things . . . are aggregates of atoms that dance and their movements produce sounds. When the rhythm changes, the sound changes as well. . . . Every atom sings its song continuously, and that sound creates shapes in every instant."[11]

Each atom sings, and that sound creates its shape; this is valid for cymatics, for the TFF, for all becoming, since nothing is excluded from the matter that is produced by sounds and music, as differentiated from the absolute silence that is the matrix. Being elastic, a sound wave moves molecules in the medium in which it is propagated (air); it does the same when the medium is water or other matter, reorganizing it into ordered structures, expressions of basic code rhythms. In other words, the noise that characterizes sound is only a shell of messages that can be translated as images. The alternation of tones and pauses (empties and solids) are numerical sequences—sequences like bar codes—that produce music or images. The secret of life is in the basic code. It is in the numbers. Pythagoras considered all numbers as sacred, divine intelligences. So also is the binary system of cybernetics. The alternation of solid and empty, one and zero, black and white—the entire universe is designed on such alternation, and everything resounds because everything produces sound.

Of these symphonies, the ear only distinguishes a minuscule band of audible sounds. Nature has selected the limited scope of sounds that we need; everything else is lost. Molecules shout, planets echo, stars clash, but we do not hear them. The pulsars are just the tip of the iceberg of the concerts performed by the celestial bodies, the music of the spheres, from the universe of galaxies to the microscopic world of particles.

Where does noise end up? Where is matter silent? Which is the smallest unit capable of producing noise? The quarks? The strings? Everything that has an identity generates noise, even the empty spaces (that are not empty). The only matter that does not produce noise is matter that has never been differentiated, that does not have individuality of shape and is lacking in identity. The only true silence is that of pure matter, as Dante described it, "where the sun is silent" (*Inferno*, I:60).

Sound Creators

Throughout the ages people have associated the creation myth with the creative act as a Verb or simply the Word. According to the ancient Egyptian doctrine of Heliopolis, it is the sun god Ra who creates the world with his divine Verb, *Hike,* which means "magic power of the word." Hike is the personification of the commanding creator.[12] In spell 261 of the *Coffin Texts,* Hike declares, "I am he whom the Lord of all made before duality had yet come into being . . . you were all created afterward, because I am Hike, the Creator of the order I live in, the Verb that will never be destroyed, in my name of spirit."[13]

In the Jewish, Christian, and Muslim traditions, God is manifested as Verb (*Verbum*), equal to the Word (*Logos*) of the Hellenistic philosophy, to which even John the Evangelist refers. In India, it is the god Krishna who creates through the sound of the flute. Even in the American and African shamanist cosmologies, sound is responsible for the genesis of the universe. The secret appears to be in the interaction between sound and matter; there is correlation between the sound and an organizing principle of matter and the field of informed forces.

There is uninterrupted direction in the background, organizing the huge mosaic that is being composed. The texts from Egyptian, Jewish,

Indian, and other civilizations describe the creative action of the basic code as already possessing information regarding form, even before it becomes mass. It could be represented as an immaterial shadow image of the physical body, with the lines defined but intangible and invisible. We can also envision the basic code as a choral work, harmonious and imperious, an astonishing concert of sounds that gives orders to the molecules and receives answers back from them. From the huge atomic holes the shouts of the particles come up, expressions of frightening whirls, until they transform into a harmonious choir, the molecules all singing. There is a continual movement of coming and going between mass and field and back again—the breathing of things.

According to the School of Heliopolis, the divine Verb activates the shape, but it does not produce a sensible reality because God must first *see* the shapes (just as in the biblical Genesis: "and He saw that it was good"). Hike, the Verb, is the intention: it confers the *possibility* of existence, having power and shape, but it does not explain things. It is the basic code: a form without substance, invisible to us, the order of the world background that prepares the shapes of the bodies and their individuality. Only after the shape is seen to be "good," that is, useful in the economy of the universe, can the virtual shape be translated into a physical body and perceived by the senses.[14] First, there is the code, and then, if the code is *good,* the body is produced. There are many possible forms that may not manifest if they are not in harmony with the overall design. It is in the "seeing if it is good" that meaning is decided.

Once it is seen to be good it is named—"The name by which Adam called every living being, that was its name" (Genesis 2:19). The ancients gave magical power to words (*nomina sunt numina,* "names are gods").[15] Giving a name means giving life; for this reason, kings and religious men were consecrated to enter a new existence with a changed name, as if they were reborn. A name can do a lot: it can create, evoke, or substitute for an object; it can cure. Giordano Bruno writes that the Egyptians knew how to evoke the intimate essence of things through writing and certain sounds. Let us listen to him in the *De Magia:*

So were the letters of the Egyptian more adequately defined hiero-glyphics, that is, holy characters; and they had at their disposal, to designate the individual things, certain images derived from natu-ral things or parts of them; these scriptures and these items the Egyptians used to capture the conversations of the gods to perform extraordinary effects.[16]

The name is sound, which gives life and creates because it possesses information of the object. Names are never by chance; they express the essence of magic, the informed nucleolus, the basic code that precedes the object, defines it, and creates it.

Water Listens to Music

Sounds can create ordered structures on water surfaces, ephemeral geometries that fade like a sand mandala dispersed by the wind. The Japanese researcher Masaru Emoto carried out experiments on water *lis-tening* to music. Distilled water was placed between two speakers that emitted music of various kinds: the water was then frozen and observed under the microscope. It seems that the crystal structures described by Emoto were formed for a few seconds during the unfreezing process, when going from −5° back to almost zero.

The music emitted from the speakers is information, which is able to change the organization of the liquid, similar to the results of experi-ments with cymatics and the TFF. Emoto noticed that different music imprints differently in the liquid, resulting in many different crystal-lizations.[17] The crystals of water that listened to Beethoven were differ-ent from those that heard Chopin, or Mozart, or a Tibetan sutra, and so on. The crystals formed by such music are almost always harmoni-ously hexagonal and attractive. But with disorderly music, such as heavy metal, the water forms intensely disordered crystal structures (fig. 8.4).

Another aspect of Emoto's experiments is also very interesting. Samples of water from all over the world were analyzed: fresh moun-tain water, holy water, rain, water from major cities, and also water from

 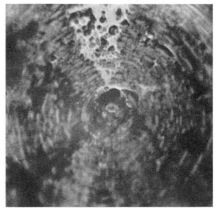

Figure 8.4. Crystals of water that "listened to" the *Sixth Symphony* of Beethoven (on the left) compared with water that "listened to" heavy metal music (on the right) (from M. Emoto)

polluted rivers. The freezing of clean water forms hexagonal, symmetrical, harmonious crystals, which is not so in the case of polluted waters. It seems that water also modifies its structure in response to pollution information.

Further, Emoto states that not only sounds but also emotions can modify matter by sending messages. Water that was daily subjected to words or mental images formed different structures depending on whether moods of love or hate were conveyed by persons sending streams of feeling to it. Receiving messages of love, water seems to answer by forming harmonious crystals, while after receiving bad feelings, it produces images under the microscope similar to those of water that "listened to" heavy metal. Ask yourself what might happen to our bodies, which are three-quarters water, after attending a music concert. In addition, what happens to our physical body when we are feeling hatred, resentment, rage, and so on?

According to Emoto, structural changes in water are observed even when a message is written on paper and applied for at least twenty-four hours to the glass containing the water. It does not matter in what language a message is written to the water, because water does not perceive words, but rather states of mind. A similar result is gained when, instead of a written message, a picture or a drawing or a symbol is applied to a glass of water for a few days, just enough time for the waves emitted from the images to inform the water. The crystallization is influenced by those images.

It seems extraordinary, but this indicates that there is a thread linking the sound waves and the mental and graphic images, for they are all able to inform water so that it crystallizes differently. What force are we talking about here? What type of waves come from a written word or a graphic image? It is not the sign, but the intention and emotion that accompany it that leave traces in the water. It seems that word, sound, sign, or image are equivalent. They convey something that is not word nor sound nor image nor sign. It is information, information from backstage that influences the world stage.

Emoto is not the first to use the forces emanating from written or mentally conceived images and signs; they have been known of and used for thousands of years. The difference is that now serious scientific investigation of their impact is possible. We are not investigating magic, but the other side of things, which—like everything else that is *other*—is frightening. Let us proceed with our exploration into the interactions between sounds and matter.

Crop Circles and Holograms

Crop circles have been mused upon by scientists, mystics, and alchemists alike. It has been known for years that in wheat fields almost everywhere in the world complex and extraordinary agroglyphs (so-called crop circles) appear. I will not talk about their origin because it has nothing to do with us now. Rather, I want to focus on how they are produced, since many scientists agree that certain sounds and infrasound may be responsible, with the help of intense heat (water in the underground water stratum also may play a role).[18]

Infrasonic vibrations between 5 and 5.2 Hz have been registered inside the crop circle formations, even hours after the onset of the crop circle; such vibrations are capable of reorganizing water. If a container of water is left in the center of a crop circle for a few hours, the water crystallizes in the same formation. According to Emoto, even if water is placed on a photograph of a crop circle for several hours, it will do the same.[19] What Giordano Bruno writes about letters and hieroglyphs is also valid for the crop circles: the whole is reproduced in its parts, like in fractals, like in a hologram.

Many factors are involved in the phenomenon of information transfer, such as sound, heat, and light. Talking of light, let us see what effects can be produced with a unidirectional and monochromatic light like a laser; then we will examine again the ultraviolets and the ultrasounds. Let us start with the laser, to understand what exactly a hologram is. Holography is a technique used to photograph objects in a three-dimensional way. Invented in 1948 by Dennis Gabor and completed in the 1960s with the discovery of the laser, it presents a goldmine of confirmation for the basic codes.

Here is the recipe for producing a hologram. Turn on a laser and make sure the light beam is separated into two. Take an object, an apple for example, and place it so that one of the two beams, after having bounced off it, will land on a photographic plate. With a system of mirrors, deviate the other light beam to make sure it collides with the first on the photographic plate without hitting the apple. The two separated beams are reunited on the plate after only one has reflected off the apple. What happens on the plate now? Do you think that the image of the apple will stay imprinted on it? No. The plate reproduces something that doesn't have a shape; it is unclear what it represents. Think about the concentric circles that are produced when you throw a handful of pebbles into a pond. The waves rapidly expand until they crash into each other and create incomprehensible intersected images.

On the plate of the hologram there is something similar, confused waves that overlap, producing alternations of thin black and gray networks. They are called interference patterns or network diffraction. They do not express the shape of the object, but they still contain its information. All that is needed for the apple to reappear is for a new beam of light to pass through the interference pattern. Then the apple reassumes its three-dimensional form, projected in space as a real fruit (fig. 8.5).

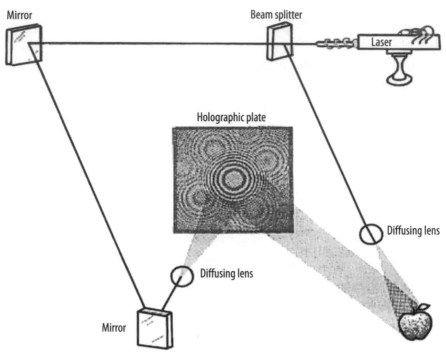

Figure 8.5. In the hologram the laser is divided into two beams: one is bounced off the object (the apple), and the other collides with the reflected light of the first on the photographic plate.
(from *The Holographic Universe,* by M. Talbot)

Holograms exhibit two very interesting characteristics. The first is that their three-dimensionality can be so convincing that an observer will believe the image really is, for example, an apple. Try touching it, however, and the apple will reveal itself for what it in fact is: empty space.[20] The other is the ability every part of the holographic film has to recreate the entire image. By putting under the laser beam even a fragment (any fragment) of the plate on which the interference pattern of the apple is imprinted, the image will be of the entire apple. Every part of the holographic plate contains all the necessary information to build the complete image (holography derives from the Greek name *olos,* which means "everything"). Crystals, for example, possess holographic properties because they can rewrite the information of the whole from every part of the crystal lattice; this is why they are used for computer memory. Fractals are also holographic.

Let us remember these properties of the hologram, because, at the end of our journey, on the other side, we will return to them when we discuss the theory of the nature of things. For now, let us return to our discussion on light and water.

Lights and Sounds

Sound and light, mostly ultraviolet, are found in all organisms and things; they control functions and are a means of communication with the environment. Water is capable of emitting light, a weak *luminescence* in the ultraviolet spectrum, with two bands around 360 nanometers and 410 nanometers.[21] This is valid for any type of water, from distilled water to water stored for months. The intensity of the luminescence varies, depending on the preservation time and the addition of trace substances. In water, luminescence seems to result from its own structuring, indicating that water is a self-organizing system.[22]

Certain sounds can free the light from water. B. P. Barber and S. J. Putterman, two researchers from Los Angeles University, in 1971 commented on their experiments on sound-luminescence, namely the emission of light by a liquid hit by a sound wave: "The duration of the sound-luminescence impulses that we observe is so brief—less than 50 picoseconds—that one wonders if some collective mechanism doesn't stimulate the molecules to emit together." What is interesting is that the emission of light behaves almost as if it is controlled by an invisible "orchestra conductor."[23] But who is the conductor? What determines the coherence of the system?

More and more people advocate the existence of some ordering principle in the organization of systems. As mentioned earlier, we have labeled these stage managers who govern the other side of things systems of intrinsic regulation, but, irrespective of what they are called, many researchers have noticed their presence for a long time. Already in 1933, Marinesco and Trillat discovered that a strong ultrasonic field can affect a photographic plate immersed in water; then it was noticed that such a phenomenon was accompanied by a weak luminescence. Other liquids (like human plasma) and even liquid metals can be made luminescent. The intensity is proportional to that of the sound wave and inversely proportional to the frequency.

Using a piezoelectric transducer,* Barber and Putterman sent a wave to water contained in a spherical jar of quartz, and it emitted a weak blue luminescence of at least 3.3 electron volts, visible to the naked eye.[24] Despite the visible "tail" of blue-violet, the peak of the spectrum was in the ultraviolet range. They also found that, stimulated by ultrasound, water emits light at regular impulses, at each cycle of the sound wave, which is most surprising.

The capability of a precise rhythm to stimulate water molecules to emit light in unison takes us back to the physics of collective phenomena and Preparata's coherence theory of electrodynamics. As he suggested, molecules can be envisioned as minuscule radio transmitters with small atomic antennas. It is possible that certain antennas with particular frequencies exchange messages. How? By putting themselves in phase: vibrating together with the electromagnetic field of thousands of messages, rapidly increasing the scope of their connection.[25] This is just like two cell phones that call each other and manage to communicate because they use the same channel of frequencies. Think about those radio alarm systems that protect houses from unwanted visitors. The central command coordinates the frequencies that the alarm sensors emit at regular intervals in order to stay in communication with each other. Using their frequency band, they talk and tell each other they are fine: they are performing their duty regularly.

As we mentioned earlier (in chapter three), when matter and field oscillate in phase, they produce high levels of consistency and coherence. According to Preparata, water has an exceptional capability of responding to the electromagnetic signals that are sent to it. The famous magnet that is the molecule of water, pulled from both sides

---

Piezoelectricity is electric polarization in a substance resulting from the application of mechanical stress. Piezoelectric substances are able to convert mechanical signals (such as sound waves) into electrical signals, and vice versa. Piezoelectric transducers are crystals, mostly quartz, that vibrate at the same frequency of a wave applied to them.

by its negative and positive poles, behaves as an electric dipole that emits frequencies while rotating. This is why water is able to talk with any other substance.

Secret Fire

We have discovered that, though it appears simple to our senses, water—"humble, and precious, and pure"*—knows how to communicate. Its chemical-physical characteristics make it capable of communicating, of recording information and then releasing it, just like when a song is recorded and then played back. Coherent electromagnetic waves (with higher-order information) manage to inform water by imprinting messages. Therapeutic information from a medicine can be transferred (TFF) or from mineral, plant, or animal substances, as in homeopathy (where the effect is inverted because of progressive dilutions with succussions). Sound waves can record complex geometric shapes on water as in cymatics, or in the form of crystals. Particular crystal structures have also been formed by currently unidentified signals, like those emitted by writing, images, or symbols.

The "dialogue" between ultrasound and water results in emissions of visible and ultraviolet light in coherent rhythms, almost intelligent. The ultrasound that frees light from the water reminds us of the Psalm: "The voice of the Lord separates the flames from the fire."[26] There is something miraculous in seeing light being freed from a body that apparently has no light in it. However, light is not only hidden in water but also in cells and in living organisms. We are facing what alchemists called the *secret fire* that vibrates in all things, which Parmenides called the daimon† and philosophers referred to as *virtue*. Is this secret fire, the flame that bursts forth, not the information of the basic codes? Isn't it the separation of the flame from the fire that transfers the information in TFF?

---

*St. Francis of Assisi, *Canticle of the Sun*
†In the fragment quoted earlier, "The narrower bands were filled with unmixed fire, and those next to them with night, and in the middle of these rushes a portion of fire. In the midst of these is the divinity (daimon) that directs the course of all things."

To this the alchemist Raymondo Lully gives us an answer by describing the secret of fire like this: "It is an instrument that is in matter and is directed to make what we seek with its movement. . . . It is directed by the *informative virtue,* which moves the matter toward form."[27] "The informative virtue, which moves the matter toward form"—this is the basic code, which guides the design of the shapes, the map to build the bodies, the control center of the combined matter in the bodies. It also provides the control and regulation system for water. But it is a structure superior to the water, with "decision-making power." To understand it more clearly, we need to leave the paths where we have found molecules singing, objects talking to each other, medicines telling stories, and music painting on water.

Now it is time to investigate the other side of *cellular* organisms; along the way we will bump into plants, animals, and living beings. Who will welcome us? I can see an onion approaching . . .

9 ⊶ COMMUNICATION BETWEEN CELLS

In the following two chapters, we will see that organisms exchange information that does not depend on sensory organs. Cells and organisms communicate through molecules and signals generated by the variations in the field. It is likely that these signals play a primary role in governing the physiology of the organisms. As in the inorganic world, cellular signals allow things to communicate with each other; they control the metabolism and rapid transmissions (in real time) between cells of any part of the organism. From top to bottom, cells are in constant radio contact with each other, and everything is always informed about everything.

We will see how the information networks involve all bodies, cellular and noncellular, and how the relationships between cells and organisms are the same as those between individuals and groups. We will see how the emotional component is a part of the biological universe and how it can influence the extrasensorial communications between plant and animal organisms. Let us start from the very beginning.

In the Beginning, There Was the Onion

In the Soviet Union in 1922, the Russian biologist Alexander Gavrilovic Gurwitsch, director of the Institute of Histology at Moscow University, was carrying out research on cell division; he was convinced that the multiplication of cells could also be stimulated by natural oscillation. To verify the hypothesis he used a simple plant substrate: onion bulbs. Later he also used onion roots in a very long series of experiments showing how the cells transmit signals, which had never been attempted before. Gurwitsch continued until reproducibility was obtained. Then he announced that plant organisms emit *mitogenetic radiations* capable of increasing the speed of mitosis (cell division).[1] In other words, they enable the cell to multiply faster.

Here is his experiment: he placed two young onion bulbs with roots attached very close to but not touching one another, perpendicularly (fig. 9.1). The horizontal one was placed only 5 mm away from the vertical one. He placed the horizontal root in a glass cylinder covered with a metallic sheath with the ends left open. The vertical root was screened in exactly the same way, except for a small window corresponding to the area made by proliferating cells, called the *meristem*. The window provided a target for the other root, which was placed so that its base pointed to the meristem. After a few hours, the onions were examined under the microscope. Gurwitsch noticed that in the target section of the vertical root with the cells were more numerous compared with the shielded cells. The incidence of mitosis was at least 25 percent higher.

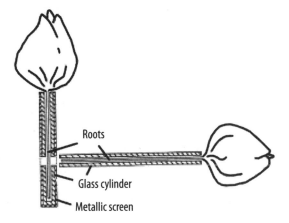

Figure 9.1. Gurwitsch's onion bulbs. The root of the onion in the horizontal position spontaneously irradiates the vertical one.

What does this mean? The tip of the root emitted something that stimulated the cellular growth of the other root, but only in the uncovered area. Gurwitsch found that this experiment could be reproduced if a quartz filter was placed between the two roots, but not if gelatin or glass was placed between them. From this he concluded that the waves radiating from the bulbs were in the ultraviolet band (UV). Since UV rays are intensively absorbed by the tissues, he thought that the mitogenetic radiation would not act directly, but by inducing the cells to a second radiation, which would be the true cause of the increased growth. Apparently it was able to transmit from one cell to another in the same organism.

The first releases published by Gurwitsch in 1923 caused such a stir that the experiments were repeated in France, Germany, England, Italy, and the Soviet Union, and (despite numerous instances of failure) many researchers confirmed the results. It was concluded that cells are able to stimulate the growth of other cells by sending very weak frequencies.

According to Gurwitsch, all organisms have sources of self-radiation, which stimulate the homeostatic mechanism that regulates the growth rates of the cells. This prevents the cells from multiplying beyond the number established by natural laws and ensures they maintain their natural intended form. But this brings us back to the lingering questions about what shapes matter and what is meant by "natural laws"? Once again, we can think of a basic code, the guardian of the boundaries of form.

Dennis Gabor (who was awarded the Nobel Prize in Physics forty years after his discovery of holography) successfully recreated the *Gurwitsch phenomenon,* together with T. Reiter, and they published their results in 1928.[2] They confirmed that embryonic tissues and malignant tumors have a high degree of irradiation, more intense in younger tissues and quickly growing. Reiter and Gabor—who were then scientists at Siemens and Halske Electric Company in Berlin—managed to modify the development of microorganisms by subjecting them to a source of ultraviolet radiation selected through a system of prisms and lenses. By exposing eggs from sea urchins to UV radiations, abnormal larvae were obtained. This also happened when they were subjected to frequencies emitted by certain organisms (for example the *Bacterium tumefaciens*).[3] The German biophysicist Rajewsky also confirmed the existence of mitogenetic radiations.[4] The Italian professor G. Cremonese in 1929 managed to imprint (after a selection for filtration) photographic plates with frequencies emitted by living organisms.[5] The variety of results suggested that these cellular emissions could belong to a range of very different frequencies.

Thus, in the twenties, something important was happening that caught the attention of the scientific community. A new frontier was outlined with the discovery that cells not only build molecules, but they also generate impulses like tiny radio stations. With the discoveries of mitogenetic waves and those able to imprint photographic plates, there was good hope of establishing some new approaches to cellular physiology. Scientists realized that to gain a correct understanding of the cell, it was necessary to explore the physics of frequencies in areas that normally had been explained only by molecular chemistry. However, it was still just the beginning of a great adventure.

In the thirties, Italian doctor Giocondo Protti, inspired by Gurwitsch's work, discovered that there is a radiating power in human blood similar to that observed in roots and plant bulbs.[6] In fact, the energy coming from a drop of blood is so intense that it can stimulate proliferation (reproduction) in yeast cultures; it can also transfer this energy to another drop of blood lacking these properties. Just as with the mitogenetic waves, the energy from blood can even go through quartz; it is therefore in the ultraviolet range (fig. 9.2).

Protti noticed that hemoradiation (that is, the emission frequencies of blood) tends to increase during pregnancy and in situations where functions are at a maximum, while it decreases during fasting and in cancer cases.[7] The Gurwitsch School observed that cancer sufferers have lost much of the radiant properties within their blood, while the tumor itself is very rich in mass, which could be the reason for the stimulation of its cell multiplication. According to Gurwitsch, cancer occurs when hematic radiation is concentrated at a single point in the organ-

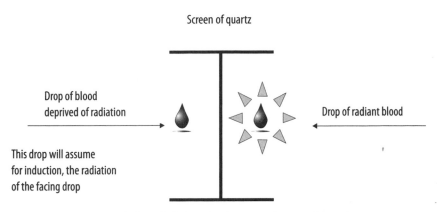

Figure 9.2. Transfer of ultraviolet hemoradiation through quartz to a drop of blood lacking that radiation (from G. Protti)

ism, which accelerates the proliferation of the cells. Protti discovered that human serum has an oncologic power, that is, it can destroy tumor cells. He also found that by increasing the radiating power of the blood, temporal cells could be killed by means of light radiation (cytophotometric).[8] Treated blood can in turn radiate and make non-radiating blood begin to radiate again. Injections of blood with a high radiating power increase the vital tone of animals whose blood has been deprived of radiating power by prolonged fasting: they brought about a reduction in tissue acidity, with obvious improvements in the course of many chronic pathologies, even degenerative ones.[9] What wonderful things came out during the years of research on the "light of the blood," as Protti defined it. And to think, all this from an onion!

And Then There Was Yeast

The Gurwitsch School also discovered that small quantities of fresh yeast added to the blood of a cancer sufferer could make it radiate once again, even after the removal of the yeast. Protti carried out many experiments with a strain of *Saccharomyces* in which the high power of radiation was able to destroy growing tumor cells. The experiment was reproduced with both glass and quartz diaphragms separating the yeast from the tumor tissue, which once again proved that these are ultraviolet waves, since the radiating power of fresh yeast is only transmitted if the diaphragm is made of quartz and not of glass (fig. 9.3).

The cancer cells are not destroyed if the yeast is dry.[10] Protti writes that the yeast, like the tumor, is also "anxious to live": it multiplies rapidly, emits intense radiations, and has a high glycolytic power and other characteristics similar to the cancer cells. The ancient homeopathic principle of *simila simillibus curentur* seems to be respected: the tumor can be cured by something similar to it. In fact, particular quartz needles filled with yeast and injected into tumor masses proved to be able to kill the malignant cells by "colliquative necrosis," compared with empty quartz needles.[11] In other words, the radiating power of the blood and certain fresh yeasts, in certain conditions, could destroy tumors. Protti measured the radiation emitted by these yeasts, with waves measuring between 1,900 and 2,400 Å; once again, we are in the ultraviolet area.

All these researchers, from Gurwitsch to Gabor and Protti, first communicated their results to the International Congress of Electro-Radio-Biology held in Venice in 1934 under the high presidency of Guglielmo Marconi. That congress was an extraordinary and unrepeatable event! Pier Luigi Ighina was also there; he was already conducting

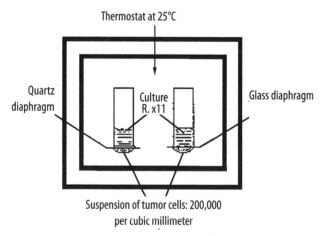

Figure 9.3. Using a setup similar to that shown here, in which quartz and glass diaphragms were both used to separate yeast cultures from suspensions of tumor cells, Protti found that the destruction due to the radiating power of fresh yeast is only transmitted if the diaphragm is made of quartz and not of glass. (From G. Protti)

experiments on the ability of information to modify matter. The days in Venice were a great event and were a clear indication that they were close to concluding something very important. But then war broke out, and the research was completely forgotten about.

Everything was resumed in the 1950s, when scientists started to talk about waves emitted by cells. A group of Italian physicists guided by Ugo Facchini from Milan resumed the experiments of Gurwitsch, who at the time was dying in Moscow, aged 80. The physicists wanted to verify the hypothesis that living matter emitted radiation, using photomultipliers to analyze the seeds of cereals and pulses. Photomultipliers are able to capture and amplify very weak fluxes of light, making them capable of measuring the weakest luminous radiations from living organisms that have been exposed to photonic stimulation. They were able to affirm that radiation from the analyzed seeds presented a band of around 500 nanometers in the optical range, with characteristic *bioluminescence* of a very weak intensity, millions of time weaker than that of a glowworm.[12] In other words, cells emit light.

In the 1960s, some tissues and other cellular cultivations were used to irradiate different biological systems, and it was confirmed that blood, yeasts, bacteria, the eggs of sea urchins, and other organisms emit and receive mitogenetic waves, amplifiable by being passed through nutritive solutions with bacteria. In the same period, the Soviets resumed their research into ultrafine luminescence. Alexandra Gurwitsch, replacing her father at the Academy of Medical Science in Moscow, together with A. S. Agaverdiyev and others, was certain that the Gurwitsch radiation exists in all forms of plant or animal life, from the simplest to the more complex, and that the spectrum and the intensity of these waves varies according to species. Particularly, they thought the emissions could increase noticeably when a biological system is at a dying point.[13]

Also in those years, in Los Angeles, the engineer George Lawrence wanted to attempt Gurwitsch experiments; for that purpose he invented a piece of high-impedance equipment with which he studied different stimulations of cells in slices of onion one-half centimeter thick and connected to an electrometer. He discovered that they reacted not just to certain irritating agents (such as smoke) but also to the mental image of their destruction. Yes—every time the scientist manifested an intention to destroy them, the cells reacted in different ways according to the psychic qualities of the person having the thought.[14] Just the scientists' imagination of dramatic events for the cell seemed sufficient to induce variations in cellular response. Intention is a very powerful form of energy, as we will also see with plants. Lawrence's experiment seems to be the first in scientific literature to present the idea that imagination, guided by a real intention, can influence cell metabolism. We will come back to this point.

Only in the 1970s did scientists finally start to recognize that such radiations are useful to the cells for communication.[15] Alexander Dubrov was the first one to note what was then called mirror image. The researcher studied the behavior of two cells in glass tubes placed next to each other; he assessed the electrical potentials and biochemical parameters, hoping to get a glimpse of new impulses in their cellular growth, but he did not discover anything new. He injected one with a poisonous substance, with the intention of observing whether the distress was transmitted to the nontreated cell, but there was still nothing. However, when he substituted quartz tubes (which allow the passage of ultraviolet waves) for the glass ones, the symptoms of distress manifested themselves in the nontreated cell. This was the first step toward a great discovery that carries the signatures of a group of Siberian researchers.

In 1972, at the Institute of Clinical and Experimental Medicine in Novosibirsk in Siberia, some researchers, inspired by the observations of Dubrov, started a long series of experiments on cell cultures. Two glass balls containing cells of the same human tissue were joined, separated only by a glass diaphragm. In the first culture, a poison was introduced to kill cells. The second, not being infected, remained intact. By repeating the experiment using a diaphragm of quartz instead of glass, the cells of the second culture suffered and died even if not infected (fig. 9.4). This happened with different types of viruses and poisons inoculated only on the first culture and without the passage of molecules from one to another. The death of the cells of the second culture

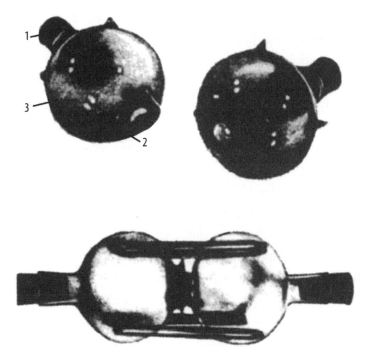

Figure 9.4. The Novosibirsk experiment: two glass balls separated by a quartz diaphragm, containing cellular cultivations. The poisoning of the cells of the first ball induces death also in the second (from F. A. Popp)

seemed to have been induced by electromagnetic signals very close to the ultraviolet band.[16]

The Novosibirsk experiment was repeated (with 80 percent reproducibility) thousands of times in the 1970s. After five thousand experiments, in 1981 the prestigious Soviet science magazine *Nauka* published the work authored by V. P. Kaznachev and L. P. Mikhailova titled "Ultraweak Radiation in Intercellular Interactions." The response to this event was so big that some risked using the term *cellular consciousness* to signify a form of control mediated by electromagnetic waves emitted by cells. In other words, this work indicated that forms of communication outside of biochemical paradigms could be possible, suggesting structures and mechanisms of action different from those previously known. The Novosibirsk experiment opened the doors to a new world of transmissions via invisible waves, all too often interpreted in a more esoteric sense than in a scientific one, with serious damage to research. Other discoveries like that of electricity, magnetism, and radioactivity experienced a similar journey.

With Novosibirsk we entered a new era. We acquired the knowledge that if we expose healthy cells to signals emitted by dying cells it can be as dangerous as exposing them directly to a virus or a chemical poison. This is the power of information in the organic world, information that was being transmitted as invisible waves. We came to understand that cells constantly emit photons and that their emissions can increase rapidly in situations of suffering. The experiments of Novosibirsk were repeated in the 1980s in many countries, in Australia,[17] Brazil,[18] China, Japan, Austria,[19] Poland, and the Soviet Union,[20] in the United States,[21] and in Germany.[22]

Some researchers[23] speak of "waves of degradation" emitted by dying cells, which appear to be caused because of the loss of their internal balance. These could stimulate other cells to enter into mitosis so that the number of cells in the organism tends to remain constant. Even if barely reproducible, these experiments are interesting because, if it is true that cells can emit contrasting signals (apoptotic and mitogenetic waves), it means that there is a homeostatic control even in small cellular groups, which presupposes that the system of intrinsic regulation could be the field connected to the cells themselves.

The Light of Things

Dying cells emit a lot more photons, a fact photographed in the Kirlian pictures of dying organisms. Surely you have seen images of leaves, fingertips, or other objects surrounded by luminous halos, like those of the saints. This is the Kirlian effect: a picture obtained by putting the object against a photographic plate that receives a high-frequency and high-voltage current.

Discovered in the nineteenth century by the Czech professor Barthélemy Navratil, electro-photography was perfected by the Polish doctor Jodko Narkiewicz, who was the first to photograph what the

Figure 9.5. Kirlian photograph of a leaf. The leaf form is clearly visible in the luminescent halo. (From L. Galateri)

local press called "electric radiation of the human body."[24] In the twenties Mr. and Mrs. Kirlian from the Soviet Union set up a device that is used today all over the world. If the intensity of the charge is sufficient, the image of the object is surrounded by a luminescence that in noncellular bodies and things is uniform and constant, while in organisms it varies in intensity and quality (fig. 9.5).

Unfortunately, the nonscientific usage of Kirlian photography by *prana* therapists, parapsychologists, and healers in order to document a mysterious halo, the so-called aura, and to quantify a person's therapeutic power, discredited Kirlian photography in the eyes of the scientific community, and the effect ending up being considered esoteric and sometimes even fraudulent.

In 1970, the Russian physicist Viktor Adamenko compared the Kirlian effect to the phenomenon of St. Elmo's fires: flames and blue flares (described also by Shakespeare in *The Tempest*) that can accompany thunderstorms when the high tension causes the air to ionize. Some people think that the Kirlian halo is the bioplasma of the photographed object; other people say it is electron beams. Adamenko discovered that in the human body, the Kirlian light is concentrated in the Chinese acupuncture points. This suggested to the German Peter Mandel the possibility of a diagnostic system based on studying the acupuncture points of the hands and feet photographed with the Kirlian method.[25]

Presman argues that the luminescence expresses a form of communication with the environment,[26] a hypothesis that is not far from ours. The studies carried out by Adamenko in the Soviet Union and by others[27]—particularly Stanley Krippner at the Maimonides Medical Center in New York[28]—have registered the following observations:

- The so-called lie detector exploits the power of the emotions to vary the conductivity of the skin: telling a lie will result in minimal variations of the measured values. The same happens with the Kirlian effect; it can vary in intensity, shape, and color when there is emotional stress. For example, the images of the fingertips of relaxed subjects are different from those of anxious ones.

- Electro-photography is so sensitive that simply touching the photographed subject is sufficient to alter the chromatic values being recorded.

- Being in a state of relaxation induced by hypnosis, meditation, or the intake of cannabis seems to increase the dimensions and luminosity of the halo, which is in turn reduced when the subject is tense.[29] The images of the fingertips may also vary due to illness, and this can be used for diagnosis.[30]

- The Kirlian effect seems to express a field of energy that surrounds and permeates the object and is sensitive to emotional moods. Could this vibrating light be information exchanges with the environment?

- Certain luminous halos certainly imply a field informed by the basic code of a thing that preserves identity and shape; for example, in the Kirlian photograph of a stemless leaf, the luminescence also includes the area of the missing stem (fig. 9.6). This is called

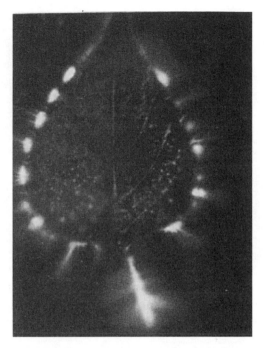

Figure 9.6. "Ghost effect": in this Kirlian photograph of a stemless leaf, the luminescence also includes the area of the missing stem. (From L. Galateri)

the *ghost effect:* the halo also surrounds what is "not there" or, better, "what cannot be seen, that was expressed as the stem." Although not perceptible by the senses, the stem is still present because the unity of the shape includes it.

Here we have an instance of shape without substance. It is plausible that the luminous image of the stem indicates that the field of the leaf—which contains the information of the entire design and image of the leaf—remains intact despite the amputation, exactly because it has to maintain the unity of the shape. Remember that, according to Harold Saxton Burr, the electric field of a bud already has the shape of the adult plant. Perhaps we can think of it as a hologram over time.

In other Kirlian photographs of plants, such as photographs of leaves with some missing parts, the luminous halo tends always to re-create the original structure (fig. 9.7).

Figure 9.7. Despite the apex of this leaf being torn, the luminescence
tends to preserve the shape of the whole leaf, as guided by
a field that maintains the shape. (From L. Galateri)

Another observation is that the halo of a leaf cut twenty-four hours
before is much less intense than that of one that has just been cut.
However, if two such leaves are photographed together, the field of the
more vital one includes the weaker one until they become one unit, as
if they put their energy together (fig. 9.8). All this data sustains the
hypothesis of a field informed by its code.

Another experiment that has been done is to place a leaf in a glass
for examination, then to remove it. If a photo is taken of the empty
glass after a few hours, a luminous halo will still be visible, just as if the
leaf were still there.[31] This is a similar phenomenon to that observed
by electro-acupuncture doctors with phials of tested homeopathic remedies: even after the content evaporated, the phial could still be used for
testing. The information of the substance was preserved in the glass as a
sort of an imprint. Is this a residue of waves or sounds that still vibrate
in the glass's ultrastructure? About the imprint left by things, Ighina
writes that he has noticed that objects leave prints of themselves; organisms leave a memory of themselves on things or in their field.

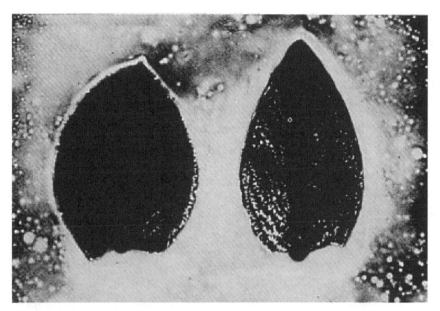

Figure 9.8. Interaction between the fields of a leaf of privet just cut
(on the right) and a dying one (on the left). The tendency
is to form a unique field. (From L. Galateri)

Another Soviet scientist, Lepeshkin, poisoned yeast with sublimate ether inside a quartz tube. When the cells died, the photons they emitted in the moment of death were sufficient to blacken a photographic emulsion solution. Lepeshkin measured the wavelength of this radiation, and it was about 2,000 Angstrom, and the caloric emission of the dying yeast was about two calories per gram.[32] A leaf taken from a tree, when examined with the Kirlian method, also showed an increase in the intensity of photonic emission shortly before dying, visually confirming what the American researcher Cleve Backster (whose research we will explore in the next chapter) identified in the twenties as the "cry of pain" of plants and other organisms at the point of death.

The Light of DNA

The work of Gurwitsch and others on the photonic signal emissions from cells was resumed in the 1970s by Fritz Albert Popp, then professor at the University of Kaiserslautern and later director at the Institute of Biophysics of the same city. In the years between 1971 and 1973, Popp carried out research on the genesis of cancer by studying a model of the isomers of benzopyrene and aromatic hydrocarbons taken from tar (isomers are two or more compounds with the same formula but with different properties due to different arrangements of the atoms in the molecule). The 3,4-benzopyrene causes cancer, the 1,2-benzopyrene does not. Unlike the 3,4-benzopyrene, the 1,2-benzopyrene does not absorb light in between the visible and the ultraviolet bands. Popp concluded that the carcinogenic effect of the 3,4-benzopyrene could be linked to its capacity of absorbing light.[33]

Popp also did research into *biophotons,* and in 1975 he proved their existence. They are tiny amounts of light that all living beings are able to store and radiate. When a photon (which is a quanta of light) imprints an atom, it arouses it, causing an electron to jump to the external orbit; when the electron returns to its previous level the photon is re-emitted by radiation (visible or not). By using sophisticated photomultipliers (fig. 9.9), Popp discovered that all cells emit this type of ultrafine electromagnetic

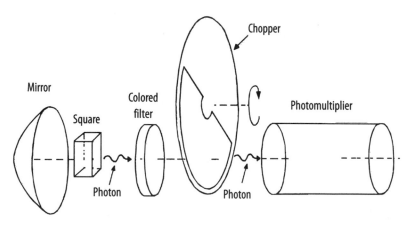

Figure 9.9. Diagram of a photomultiplier. In the darkroom, the little curve of the quartz containing the cells placed in front of a mirror is focused by a photomultiplier. The colored filter allows the selection of different colors of light. The chopper is a dark-light rotating disk that makes it possible to select the photonic signals actually emitted by the cells, distinguishing them from the background noise. (From F. A. Popp)

radiation in the range between ultraviolet and long radio waves. It is manifested in the visible as a very weak luminescence, whose intensity is 10^{18}
times weaker than daylight, comparable—if you want—to a lighted candle
seen from twelve miles away. This is not to be confused with the bioluminescence emitted by fireflies or certain species of fish or bacteria.[34]

It is worth remembering that the luminescent properties of certain
substances that emit electromagnetic waves (infrared, ultraviolet, X
rays, or gamma waves) come from energy absorbed as electromagnetic
radiation. If the phenomenon lasts less than eight to ten seconds it is
described as fluorescence; if it lasts longer then it is phosphorescence.

Popp observed that stress, disease, and suffering are accompanied
with increased biophotons. Initially he interpreted the bioluminescence
as deriving from cell metabolism, something like the sparkles that
come out of the electric wire of a tram. Later one of his students called
Rattemeyer thought that the biophotons came from the cells' DNA, the
controller of all of the cells' vital information. He was proved right. In
1981 Popp and Rattemeyer published their conclusion that biophotons
are emitted (though not exclusively) from cellular DNA and that, during mitosis, when the chromatin (a combination of nucleic acids, mostly
DNA, and protein found in the cellular nucleus) is dissolved, that is,
when it passes from phase G_0 to phase G_1, cells release massive photon
emissions. At that point the bioluminescence was not anomalous, but
formed signals of great significance to the cells.[35]

Then it was understood that biophotons have a guiding role in the
regulation of all cellular physiological functions; by exciting the electron layers of the molecules, they are responsible for the initiation of
biochemical reactions. In other words, the orders are given by the DNA,
but it is the biophotons that carry the messages and ignite the metabolic
fuse. Once again, it is possible that molecular events are the consequence
of physical reactions that are based on the energy quanta. Without the
ongoing work of the biophotons, cellular metabolism would end up in
chaos and would take life with it. Popp's studies suggest that the intrinsic
mechanism of existence must be sought in physical and chemical models that describe effects, not causes. Popp was convinced that biophotons

were emitted by the heterochromatin, part of the DNA (about 98 percent of the molecule) that is not expressed in genes and seems meaningless. It could be, instead, the most important part of the DNA, storing light to regulate the cellular biochemical reactions.[36]

Popp noticed that this is possible only if the biophotons constitute an electromagnetic field of high coherence. In fact, the more numerous the frequencies are, the greater the degree of information and the lower the possibility that other photons, even intense ones, can interfere with its processes. A weak radiation but with a high informative order will never be disturbed by a stronger radiation that is lacking order. The cell always tries to defend itself from chaotic external radiations that could confuse its communications. Nature makes cells resistant to great electromagnetic impulses, but they are susceptible to others, even very weak ones, if they are coherent, like those of homeopathic oscillations or other techniques such as electro-acupuncture and the acupuncture itself.

According to Popp, DNA behaves as a photoaccumulator: it is activated by absorbing light. It then immediately emits it in the form of coherent photons that act like laser rays. As mentioned earlier, in a laser all the waves oscillate in phase, making its light monochromatic and extremely pure. Unlike the light of a bulb that radiates in all directions (chaotic), a laser light becomes a unique unidirectional beam (coherent). Once again consistency is the key word for understanding how very weak intensity is sufficient to govern complex cellular functions. The energy absorbed by the DNA is compensated for by ultrafine cellular radiation. The DNA thus works like a radio station that regulates not only the cellular biochemical processes but also the communications between cells.

Let us try to gather these results together: according to Popp DNA acts like a radio station; Preparata and Del Giudice see molecules as antennas that emit and receive signals; and the apparatus of the TFF works like a radio station where medicines emit signals. Together they indicate that the human body and the world of things are governed by radio transmissions, which build a well-known network of communications. Prigogine wrote that it is communication that elevates life above chaos, enabling it to evolve. Cells are open systems, dissipative structures that communicate by means of countless light beams that originate from each nucleus. Even more than the membranes, it is photonic resonance that unites the cells. This has consequences for the phenomenon of recognition, immune response, and susceptibility to tumor degeneration. It is obvious that biophotons are responsible for the communication between the cells; the chemical transmitters are only the instruments.

In 1982 at the Max Planck Institute of Astronomy in Heidelberg, Popp and Beetz made the biophotons visible using a residual light amplifier with a fluorescent screen.[37] By first using cress and then another type of cell more sensitive to photons (the unicellular green algae, *Acetabularia acetabulum*), Popp managed to demonstrate that the photonic emissions periodically fluctuate according to cellular biorhythms; they are intense when the cells are developing more rapidly, and they disappear at death. In accordance with Mr. and Mrs. Kirlian and the researchers from Novosibirsk, Popp noticed an increase of luminescence in the case of strong cellular suffering: poisoning of a cell increased its photonic production prior to death, which was reproduced on the computer screen as luminous explosions and as spikes on the graph showing emissions (fig. 9.10).

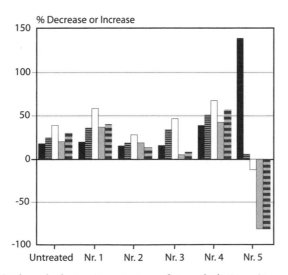

Figure 9.10. Reduced photonic emission of *Acetabularia* poisoned with atrazine (Nr. 5) compared with the nontreated control, a clear expression of cellular suffering preceded by a peak of strong photonic emission ("the cry of pain").

Popp was the first to demonstrate that two cells communicate with each other at short distances through coherent light emitted by DNA, which also controls the functions of cells. In fact, he showed that if an *Acetabularia* in a bowl of quartz in the darkroom of a photomultiplier is stimulated with white or colored light for a few seconds, its luminous emissions will appear on the screen in the shape of pulsating explosions. After a little time the intensity of light weakens, but if we put seaweed emitting similar luminous signals near it for a second, the first will be reactivated with new emissions of light; the cells seem to *dialogue* with each other. The time used by the cell to release the absorbed light from its accumulator (the DNA) is important. The more rapidly it is emitted the less good is the accumulator.[38]

DNA represents the priming and control system for cellular biochemical functions, it plays a role in maintaining homeostasis (the ability of a system to self-regulate itself to maintain its balance), and it is responsible for communications, even at great distances between cells. Through the light protected in its deep spaces, the DNA transmits information at a very high speed throughout a whole organism. This would explain the reason why the majority of the biochemical reactions in a cell take place at least a billion times quicker compared with the same systems in vitro. The biophotons are far more numerous in living beings than in in vitro systems, confirming the presence of informative networks operating in vivo; the network's effectiveness increases with the complexity of the organism.[39] This explains why in the more organized systems the response to stimulations, as for example to the TFF, is more rapid and effective, according to the Kaiserslautern postulate (that the more complex the system, the more complete the decoding of the signals).

It has been observed that there is a balance between photons and growth: when cells develop, many photons will curb proliferation, while growth increases when the photons are reduced. It is possible that in cancers this regulation is disturbed if carcinogens damage the DNA or interfere with the coherence of the emitted light. Popp thinks that cancer can be associated with (and may be caused by) the loss of coherence of the radiation field with uncontrolled cellular pro-liferation.[40] In general, each disease may be linked to perturbations of the cellular fields; perhaps the accumulation of disordered oscillations within the organism produces distorted regulations. The theory of the biophotons shows that weak impulses, well below the threshold of perception, have no value for their energy content but have an important significance for information conveyance. This points to the importance of an electromagnetic type of system of intrinsic regulation in organisms; it may be that every chemical reaction is preceded by something physical.

Water, Cells, and Light

At this point, we have to go back to water and its relationship with the light frequencies emanating from the DNA. Among its many functions, water grants stability to the DNA: if the water percentage around the double helix falls below 30 percent, it results in the denaturation (structural change in macromolecules caused by extreme conditions) of the nucleic acids.[41] Biological systems, being open, have a high degree of order because, contrary to the closed systems, they are far from the thermodynamic balance. This allows them to exchange a lot of energy and produce coherent behaviors. Life needs to be as far as possible from the thermodynamic balance.

If we warm liquid from below, its molecules get agitated in a chaotic way, but a certain amount of heat triggers convection currents that become more and more stable until they form so-called Benard cells: ordered microscopic structures like little cells in rows composing a mosaic of a dissipative type. In other words, if we move away from the thermodynamic balance (for example, with a boiling action), beyond the chaos, the molecules restructure in an orderly manner. A system can be ordered only when it intersects with a great flux of energy, which must not stay there for long: an energetic jam would destroy the coherence.

Frölich proposed the model of *coherent oscillations*.[42] Water molecules are powerful magnets (we have already seen that they have high dipole values), and they tend to couple their oscillations under the influence of an electromagnetic field: they start to dance all together, like in a ballet. In this way, energy is transmitted to the entire system, as in the model of dissipative structures. At certain stages of field intensity the molecules start to oscillate in a coherent way, emanating a particular wave that behaves like a laser. Think about it: a laser produced by water! These are *long-range interactions,* able to control the entire system. Thanks to them, the molecules inside of cells are able to interact instantaneously. And it is also thanks to these laser beams that water can receive information and release it, as in homeopathy, in the TFF, or in other methods of water activation.

Biophotons are cellular messengers, waves with high coherence that rightly fall within the models of Prigogine and Frölich. They are not information, but they are carriers. They are not generated by the cell; they are light captured from the environment, which is then emitted in a coherent way in order to transport information. The biophotons suggest a regulatory control principle in the cell identifiable with the field itself. But let us leave the cells and start to explore the complex organisms and their non-sensorial communications. At this level of organization there is a new element we have to take into account: the emotions.

10 •—• PLANT AND ANIMAL COMMUNICATION

Plant Sensibility

At New York Police Academy in 1966, Cleve Backster taught the use of the lie detector, a form of polygraph* that, when connected to a person's skin, measures the electrical potential of the body and the vibrations stimulated by thoughts and emotions. When the person tells a lie or experiences the mental images and emotional stress common under threat, electric currents are generated that change the graph.

One evening on his way home, Backster had the idea of connecting the electrodes of the polygraph to the leaves of one of his plants, a *dragon tree*. He just did it without thinking about it, moved by an irrational and apparently casual impulse that would change his entire life and deepen our knowledge of the universe. Nobody before him had thought about doing this. He discovered that the leaves of the plant, connected to the polygraph, emitted electric currents just like human beings do, with an identical graph. Since it was impossible to interrogate the plant to discover lies, Backster tried watering it. The graph

---

*A *polygraph* is an instrument able to simultaneously record biological signals and to transcribe them in graphs. The electrocardiograph, the electroencephalograph, the electromyograph, and so on are examples of polygraphs.

of the dragon tree slightly oscillated, as would a person who had been emotionally stimulated. Then Backster had the idea of burning a leaf in order to induce a stress reaction, but before he could try it, just the act of thinking about it triggered an alarmed response in the plant. It produced a pattern similar to that of an intense reaction in a human subject: a jump toward the top of the graph, the sign of typical anxiety reaction (fig. 10.1).

Backster tried to reproduce the experiment and discovered that the recorded graphs of plant reactions were just like those of human beings and that plants can perceive a signal emitted by a person before any action is taken. When he thought about burning the leaf without a real intention to do so, there was no reaction from the plant. In this way he came to understand plants respond only to real intentions, which emit emotional vibrations perceived by the plant.

Backster's work was rejected. The press was skeptical of his publications, and they joked about plants reading thoughts. However, the results were not because of telepathy but a resonance between fields. Since harmful thoughts did not induce a response unless accompanied by intention, it is evident that the plant perceived the variation of the

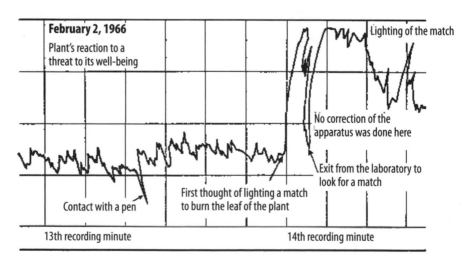

Figure 10.1. This graph was recorded by Backster February 2, 1966: the dragon tree reacted in the precise instant when Backster thought about burning one of its leaves.

field caused by the intention. Plants are not able to see a match being lit; instead they perceive the intention and the emotional variation that accompanies it.

Backster later discovered that neither a Faraday cage nor a lead shield can block this kind of communication. He also measured whether the phenomenon would still occur when there was a great distance between him and the plant: it did even when he was in other rooms or houses. Once the harmony between plant and person was established, communication was possible even at a distance of miles. He also learned that in absence of a *feeling* connection with the plant the experiment was barely reproducible, even at short distances. If there is indifference the emotional currents are blocked. Person-plant communications are stronger with certain emotional states or with certain people, like those with so-called green thumbs, an apt expression of perfect resonance.

Backster reported a case of a lack of reaction from the plants to the usual stimulations when a physiologist who took plant cuttings witnessed the experiments. In the field of this woman were imprinted violent emotions that "paralyzed" the plants; they started reacting only three-quarters of an hour after the physiologist had left. Backster interpreted that silence as a "loss of conscience" or "fainting" as a defense against potential danger.[1]

Afterward Backster observed that the arenas of the plants' reactions were not easily defined; some emotional episodes happening only three yards away were not registered, but at seventy yards they were. The plants responded not only to the aggressive intentions of human beings but also to unexpressed threats like the jumping of a dog or someone who did not love them; and they even resonated with the actions of a spider trying to avoid human interference.[2]

To demonstrate the transfer of information from the field of a person to that of a plant, Backster collaborated with a journalist who was asked about the year of his birth, which was recorded. Then, in the presence of the plant, Backster proposed a series of dates as answers. The journalist always said no, therefore lying when the right date was given as an option. The philodendron connected to the galvanometer recorded the precise moment of the journalist's lie. The experiment, reported in the *Reader's Digest,* suggests that a plant can respond to the lies of a man with the same reaction that would be recorded by a galvanometer connected to his skin.[3] This is consistent with the hypothesis of a network of universal communication.

Backster noticed traces of "suffering" in plants when some yogurt bacteria died, which suggested to him a new series of experiments leading to his famous experiment with plants and shrimp. Here it is: a device would randomly tip live shrimp into a pot of boiling water, without the knowledge of the researcher. In three different rooms there were plants connected to galvanometers. Results were recorded with and without shrimp and by a galvanometer not connected to any plant to exclude environmental interferences and current variations. The result was that all the plants reacted intensely and synchronously to the death of the shrimp.[4] From this observation arose Backster's conclusion that, during physical suffering and at the time of death, not only bacteria and crustaceans but all life-forms emit a cry of pain—as he called it—that other living forms can register. More than a "cry," it is perhaps a sudden variation of field intensity, which galvanometers, photomultipliers, and electro-photography are capable of registering.

The shrimp experiment yielded Backster fame and fortune to carry on with his research with more refined equipment, such as cardiographs and electroencephalographs. He confirmed that a plant will react violently to the breaking of a nonfertilized hen's egg and that this could be extended to the animal kingdom, since an egg connected to a cardiograph recorded the death of an egg thrown into boiling water. Human sperm in a test tube responded to physical suffering experienced by its donor at a distance of more than ten yards. Backster reported the result of this experiment in 1975 at the May Lectures in London. On that occasion he also reported that the skepticism of an individual present at the experiments might adversely affect the results, just as trusting behavior would act as a catalyst for the reaction investigated. It seems that human beings can interfere with what they observe. If a person, or any other being, unconsciously gives emotional signals that interact with the environment, the communication system is very sensitive to interferences from other waves capable of catalyzing or curbing it.

Other researchers carried out the experiments on plants also. Pierre Sauvin, an electro-technician from New Jersey, discovered that a person is able to communicate with plants at a distance, even by just remembering a dramatic event. As the person recalled the event, the plant registered a change in the person's emotional field (the space in which emotions are transmitted) similar to that caused by the event itself. In other words, memory or imagination as well as reality appear to induce measurable field effects.

Sauvin documented improved communications with plants with which he had particularly attentive relationships. He confirmed the responses of plants to all types of emotions: suffering, pain, shock, pleasure, as well as the death of cells in the environment. Tired of exposing the plants to pain in order to measure their responses, Sauvin investigated how they reacted to one of his sexual orgasms more than sixty miles away: the reaction was prompt and synchronous with the moment of pleasure.[5]

Eldon Byrd, a laboratory analyst at Silver Spring, Maryland, confirmed both Backster and Sauvin's studies, and he demonstrated on TV the reaction of a plant to different stimulations, including the intention of burning it.[6] Ken Hashimoto, from Japan, a doctor in philosophy and an electronic technician near Yokohama, then managing director and head of research of Fuji, even went as far as "talking" to his cactus. Before this he had had a role with the Japanese police similar to Backster's and had read Backster's reports. Hashimoto decided to try to "listen" to his plants by transforming the polygraph's movements into sound modulations, thus giving them a "voice." His first experiments were unsuccessful, but when his wife, who had a green thumb, gave love to the cactus, the plant reacted immediately. The sound that came from the apparatus was similar to the hum of high-voltage electrical wires, with varying rhythms and tones, which became an almost pleasant song, sometimes excited and sometimes happy.[7] In his book on the fourth dimension, *Mystery of the Fourth World*, Hashimoto declared

that he was convinced of the existence of a world beyond the one perceived by our physical senses, a dimension of which our sensible world
is merely a shadow. That brings us back to Plato . . . and the other side
of things.

The most famous follower of Backster was Marcel Vogel, a
California chemist from Los Gatos who improved the conductivity of
the polygraph electrodes by spreading a special gelatin-based agar-agar
jelly on the leaves. Vogel started his research in 1971 by trying to shower
his philodendron with love and recording the results. The researcher
compared the phenomenon to the nonverbal communication that, for
example, occurs between two people in love, or when a conference
speaker or an actor senses hostility or benevolence from the audience.
Such hostility can cause an actor to have a block, similar to Backster's
plants in the presence of the physiologist. Vogel argues for the existence
of a *vital force* pervading all living beings, plant or animal, facilitating
nonverbal communications between the different forms of life.[8] Vogel
placed himself in front of his plant, connected to the galvanometer,
relaxed, and touched the leaves while giving it affectionate emotions, as
to a friend. The plant answered with a series of escalating oscillations
that stopped after a few minutes, regardless of whether Vogel continued
emotive manifestations. It was as if the plant had unloaded its energies
and had to recharge itself.[9] In another experiment he connected two
plants to the same galvanometer and cut a leaf from the first plant; he
noticed that the second one reacted only if Vogel gave it attention: in
essence, he had to be connected to the plant to achieve results.[10] During
a conference he said:

> It is a fact: man can and does communicate with plant life. Plants
> are living objects, sensitive and rooted in space. They may be blind,
> deaf, and dumb in the human sense, but there is no doubt in my
> mind that they are extremely sensitive instruments for measuring
> human emotions. They radiate energy and forces beneficial to man.
> And you can feel these forces! They feed the human energy field,
> which in turn regenerates the plant.[11]

A psychic, a friend of Vogel, took two leaves from a saxifrage plant: on one she lavished love every day, while the other was abandoned. A month later, the second one was withered and almost rotting, while the other was green and vital as if it had just been cut. The experiment was repeated and reproduced by Vogel.[12] It seems that our emotional fields transfer information of love and health that affects matter. Vogel and Backster anticipated by years the Emoto experiences with the crystallization of water subjected to emotional states like love and hate.

Under the microscope Vogel studied the subtle dynamics of liquid crystals; he concluded that the crystals are driven to a solid or liquid state by preforms (or ghost images) made of pure energy, which precede the formation of the solid or liquids. In this he anticipated the idea of the basic code and its informed field as a map for the construction of a body.

The first experiments conducted on plants in the Soviet Union date back to the 1970s. At the Academy of Agricultural Sciences in Moscow distress signals emitted by barley sprouts at death were recorded. At the State University of Alma Ata in Kazan, using the Pavlovian technique of conditioned reflexes, a short-term memory phenomena was observed in plants: stimulated with an electroshock whenever it touched a certain mineral, a plant learned to recognize it and react to its presence, even without the electric shock.[13] In 1972, Vladimir Soloukhin, of the Akademgorodok Research Center in Siberia, published in the *Nauka i Zhizn* journal his findings that plants seem to collect ideas and preserve them for a certain time. The plants manifested reactions of fear in the presence of people who tortured them, while they expressed a smooth graph in the company of people who took care of them. Studies on the *Backster effect* continued in several parts of the world, and they all confirmed that plants (as well as any other living being) perceive signals of an emotional nature emitted by others. Feelings and emotions within the body can be captured by other beings, even at great distances. Cellular death especially may be experienced with dramatic intensity.

To summarize: human emotions produce field variations that plants resonate with; they react by modifying their own field and emitting certain currents. If there is an intimate relationship between plant and person, emotional communications can be immediate, even at great distances. Plant fields also can resonate with other animal species and respond to suffering and death. Finally, imagination produces identical effects on the plants within our emotional field as those caused by real events.

Emotions and Animals

It is a common observation that animals respond to strong feelings in humans. If a child is afraid of a dog, the variation in the child's emotional field in the presence of the dog is transmitted to the animal, which will raise its face and ears in the child's direction. Since the 1980s the British biologist Rupert Sheldrake has been studying the behavior of animals and their powers of extrasensory communication. Many species are able to grasp changes in the emotions emitted by human beings with whom they have established particular relationships or with animals of their own group. It happens in herds, flocks of birds, insect groups, and fish schools, providing that there is an affinity of belonging to the group or a particular emotional resonance, like between master and animal.

Cats and dogs react to the intention of their owner to return home, just as plants do to the intention to do harm, their reactions synchronized to the emotional wave emitted by the person. Sheldrake used closed-circuit cameras to document the exact time that a dog's behavior (such as approaching the door or watching from a window) indicated that it sensed the imminent arrival of its owner on the doorstep; even if the owner came home at varying times, the dog would sense the precise moment the owner would arrive every time. Sometimes behavioral changes would occur at the time that the owner activated the intention to return, even from quite a distance. The increased occurrence of such behaviors was very pronounced and statistically significant.[14]

These are field interferences, informational exchanges between person and animal, even at great distances. If a person releases a certain emotional wave it can influence the behavior of his or her animal. For example, when a dog's owner thought about sausages or biscuits, the dog felt the thought and appeared in front of the owner: it was not responding to the word, but the change in the person's emotional field. Just like when plants perceived Backster's intentions, animals are able to pick up emotions and mental images transmitted by humans. There is also clear evidence of communication between animals, such as a dog perceiving the death of dog friend at a distance. I cite two interesting experiments reported by Sheldrake.

The first was conducted on dogs by a psychiatrist of the Rockland State Hospital in New York. A mother and her puppy were separated, then the puppy was threatened. At the precise moment when the puppy was threatened, the mother reacted as if she had been threatened herself, even at a distance. In the same way, in another experiment, the heartbeat of a dog connected to an electrocardiograph increased at the very moment that his owner was threatened.[15]

In the second experiment, French researcher René Peoc'h compared the behavior of pairs of rabbits from the same brood, bred together for months, and other rabbits bred separately. Each animal was in a confined case, acoustically and electromagnetically isolated; the stress they underwent was assessed by the measuring of blood flow. It was noticed that when one of a pair of rabbits bred together received a shock, the other manifested tension immediately. This did not happen between pairs that had been bred separately.[16]

Animals establish resonances even with the places where they live, which allows them to find their way home again without making mistakes. The memory of a place is a great mystery: how can the migratory species, such as insects, birds, turtles, and so on, return to the same place? Among the monarch butterflies that go to winter in Mexico, it takes from three to five generations to complete the journey. Even though no individual can experience the entire journey, they seem to be provided with an innate program containing the memory of the route. In the 1950s the Dutch biologist A. C. Perdeck showed that by forcing the youngsters of a migratory species to change path, new patterns of migration could be established in a single generation.[17] Some flocks that migrated from Holland to England were captured and taken to Switzerland. From there, the adults found their way back to England, while the youngsters carried on with a path parallel to the one they would have followed from Holland and ended up in Spain. An internal program that gave them direction guided them. Some of these returned to their country of origin even if they started from a new place, and the following year they went back to Spain (fig. 10.2). In such cases, if different information cancels out the previous, it is not due to genetic mutation but to resonances between fields.

Finally, there are relationships between animals that support the organization of complex societies (such as bees, termites, and ants) or that direct complex action groups such as fish schools or bird flocks.

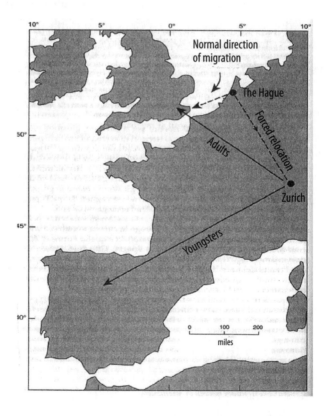

Figure 10.2. The flocks transported from Holland to Switzerland took two different directions: the adults went back to England, while the youngsters flew a parallel route, ending up in Spain.

It is as if there were a group soul or, according to Sheldrake, a *morphic field* serving as custodian of the form. Let us take the termites who build nests several yards high with underground chambers and tunnels, a network of links that require the labor of generations of workers who must interact in a precise and ordered way. It is not yet understood how these insects communicate for such lengthy periods. Sheldrake has suggested the existence of a social field containing the building plan, just like the basic code and its informed field. The construction of the termite mound depends on its morphic field, just like the orientation of iron fragments around a magnet depends on its magnetic field. When the termites build arches, they make columns first, and then they bend them toward each other until they meet. How they do it is incomprehensible, since they cannot even see. They are not guided by sight, sound, or smell: the termites know the way, and they obey an invisible design.[18] Who owns the designs of these groups of animals?

The German naturalist Günther Becker demonstrated that termites are guided by a biofield, which can penetrate material barriers, just like a magnetic field. The insects have an awareness of the whole territory and its borders. In their mounds the vertical tunnels are only built up the peripheral walls; this happens also in experiments where they are divided in several containers (fig. 10.3). In a line of vertical square

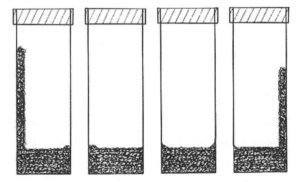

Figure 10.3. The building of vertical tunnels by the termites in captivity inside plastic containers is only on the external walls, not on the walls adjacent to other containers. The barriers of the internal walls do not shield the field that is transmitted from one container to another.

containers, vertical tunnels were constructed only in the two external containers (not on the walls adjacent to other containers) to delineate the beginning and end of the nest.[19] What gives the whole its concept? A unified field produced by the termites, which dominates and orders them so that they behave like a single organism.

This type of field was postulated in the 1920s by the South African naturalist Eugene Marais, who conducted experiments on the ability of termites to repair any breakages in their mounds. That was during the same prolific years of research when Gurwitsch discovered mitogenetic waves, the morphogenetic fields of plants were described, and it was realized that plants have fields equivalent to those of animal groups. Without contact between them (just like the researchers who were getting the same results without knowing about the others: also due to morphic waves!), even with a steel diaphragm in the middle, insects will construct the two sides of their structures identically,[20] just as if there was a predetermined plan, an organization field connected to the queen.

Similarly, the field of a flock adjusts and regulates the field of each individual so the group behaves as a single unit. The individuals cancel each other out in the unit, and this makes possible the perfect synchronicity of lightning-quick maneuvering of hundreds of fish or birds where each individual at every moment knows in which direction its neighbors will move. In the sudden flight reaction of fish, in which they seem to disappear instantaneously (called flash expansion), every individual shoots away from its place at a speed of ten to twenty times its length per second. In the case of birds, the very rapid movement occurs along waves propagating at the speed of fifteen milliseconds between the individuals. Every single movement of the whole school or flock is perfectly coordinated with the others of the same group, and therefore it is a single entity, a *super-entity* able to direct every particular aspect of its components.[21]

Something similar happens with humans. Psychoanalysts call it a *group soul*, referred to in so-called group analysis, where the group is not considered as a series of individuals (for example, people at a bus stop) but as a unique reality functioning according to what psychology calls the collective unconscious, a group field. Among the participants, there are particular relationships that result in a homogeneity of frequencies; it is clear that certain moods are often present in all members of group at the same time. Some memories may emerge collectively, as well as certain aspects of amnesia that can suddenly occur across the entire group. Such principles provided a base, for example, for therapies such as "familiar constellations," evolved by Bert Hellinger.[22] In these cases, the group soul enables the individual to interpret other people as being like himself; even without knowing one another they are able to read character, memories, and secrets as if they were already "magically" known. This resonance phenomenon can be attributed to the forces of a collective morphic field.

Communication between Human Beings

Extrasensory human communication is possible; there is extensive literature on this subject, but I will not delve into this now. I just want to point out that it is more common among people living in close contact with nature, uncontaminated by technologies. When analogical thought has not been subjected to the rigid logical paradigm, certain people are more able to accept experiences beyond the barrier of the senses; I refer to people described as primitive. Examples of telepathic communication have been noted among many African, Far Eastern, and American native peoples. The bushmen, for example, know exactly when the hunters of their tribe have killed an antelope, even at a distance of six miles, and they know in advance when they will return.[23]

When I was in the Peruvian Andes, I met the Q'eros, the magicians of the Andes. The Q'eros represent a tiny nation of the pure Inca race who for centuries have been living in inaccessible valleys at altitudes of almost twenty thousand feet. I will quote one of the many episodes I had the chance to witness. Our group was in the Sacred Valley in the company of the spiritual chief of the Q'ero community, Don Mariano, when one of the guides, Raoul, had to go home to Cuzco. He told us that he would return in time for dinner. Our hut in the forest at thirteen thousand feet did not have telephones, and at that time there were no cell phones, especially in the Andes. Something must have happened, as Raoul did not come back and was absent all night. The following morning, as we were all worried, someone asked Don Mariano to ask Apu, the Great Spirit, about the fate of our friend. Isolating himself, the Q'ero asked for news from Apu (maybe he "tuned" his field with Raoul's), and he announced that there was nothing to worry about because our friend had gone to the hospital the previous day with his sick mother, but she was now better, and Raoul would return in the evening. In fact we saw him just before sunset, and he apologized for the delay caused due to the hypertension of his mother, who had by then returned home from the hospital. If Don Mariano did "read" Raoul's field, it would mean that emotions can be written as on a board or can be imprinted as on a photographic plate. They do not reach the threshold of consciousness, but they can be "read" by those able to perceive them, especially those who have genetic and emotional links.

For more advanced Incas all this is normal: how otherwise could they communicate when they live weeks of walking distance away from the cities? People like these have developed faculties to resonate with each other at great distances in order to communicate (and without computers).

The more advanced Celts of the Scottish Highlands may also have had visions of approaching people who would appear only later. In Norway, the same phenomenon is called *vardøger,* literally "perceiving spirit," and once again it is a perception of intentions. The wife can "feel" the husband coming home as soon as she is alerted by the vardøger, and she immediately puts the kettle on for tea.[24] They call it "the spirit." How often has the field been considered spirit and field resonance attributed to divine forces! What did the Renaissance neo-Platonists call the field? Spirits! Magic is the physics of the fields.

Some people use thoughts or emotional currents to communicate with their animals while the animals sleep. They imagine themselves giving orders to their dog, for example, and the dog will carry these out once awake, like a posthypnotic command. Vladimir Bechterev, a Soviet scientist in the 1920s, studied many phenomena of orders given mentally to dogs, something that can be efficient also at great distances.[25] This is similar to a technique used by the American Indians called whispering to horses, or the vast range of experiences collected by Carlos Castaneda about his apprenticeship with Don Juan and the sorcerers of Central America.[26]

Among All There Is Always One

Fish schools and bird flocks behave like an organism formed by millions of cells. The organism realizes the most perfect harmony of the whole, so that the collective interest prevails over the individual. A liver does what it must because it is a liver, it doesn't do what every single liver cell would like (this surely reflects the collective will in physiology, but not in pathology). Liver cells vanish as single entities and independent units, and only the liver as a whole exists.

Each human being is an example of how billions of cells cease to be individuals to exist in one entity called a person. Otherwise every cell would go on its own: the liver cells on one side, the pancreas on another, and the thyroid would leave to grow in other forms. For every system there is a single direction, which in turn fails if the system is reduced into separate units. In a brick wall, the bricks are not bricks anymore; they lose individuality, and the remaining entity is the wall. The brick blends into the wall, as the cell into the organ, the fish into the school, the bird into the flock, or the hydrogen and the oxygen into the water molecule. They disappear from the world, to reappear again as an entity.

The fundamental law, from the micro- to the macrocosmic, is that the field of a group takes precedence over that of an individual. In the world of particles a field that organizes structure and identity governs each atom. The atoms sacrifice this identity by grouping in molecules. The field of the molecule (which precedes the molecule) prevails over the atoms and incorporates them into itself, just like the builder who takes identity from the bricks to give it to the wall. When molecules are organized in aggregates (proteins, viruses, crystals, and so on), the ordering fields of these structures prevail over the molecules. The same applies for those that make up cells and those absorbed in more complex systems, such as organs and bodies that can be structured in groups.

Individuals suspend their individuality if they are called on to form an organized system: it is a universal rule that the superior entity prevails over the inferior one. The cell is lost in the organ, as the fish in the school, or the bird in the flock; to form a whole is their prime function. The function is the system of intrinsic regulation. If the unit of the whole was lost, it would be a catastrophe: uncontrolled cell growth might lead to neoplasm, just as in a social rebellion one man from a whole society can cause crime and subversion to the order.

Belonging to a group promotes extrasensory communication in higher organisms as well as in single cells: a sperm emits waves of pain when its donor is harmed; the dying cells of Novosibirsk induce death in their companions; rabbits are able to transmit their shock to their partners. It is possible that resonance between fields often happens along with normal sensory communications, and thus goes unnoticed because it is masked, just as with medicines when the action of the field is concomitant to the molecular one. As a result, field resonance may be more readily observed in distant phenomenon where the senses do not reach.

Field actions have fundamental roles in an individual's survival, as is proved by the studies on animal behavior. In the course of evolution, to communicate at a distance, humans have invented the radio, the telephone, the fax machine, and electronic mail, but this has caused more natural communicative properties to fossilize. When an animal and his owner *feel* each other at great distances it is as if there is a thread between them, just like two antennas synchronized on the same frequency; they can find each other among billions of other frequencies. From these observations, Sheldrake's theory of morphic fields was developed. It is the same theory as the superradiance theory: the field generated by particle oscillations regulates them by homeostatic mechanism, enabling the molecules to connect in long-range interactions.

Quantum information, wave physics, and field communications express the thought that through the past century we have been moving from a world of solids and masses to a world of emptiness and information:

there is nothing molecular in Internet communication. Theoretical models have been affected, and there are certain scenarios where notions of molecular interactions are being replaced by the hypothesis of field interactions. Superradiance provides us with a model whereby each cell is equipped with a supervisory field that controls the system and governs the vital processes. We can predict the same for every organism and more generally for every arrangement. With an incessant self-regulation, the superior field maintains order in those belonging to its ensemble, which in turn contributes to maintaining the complex field. We can apply the models of superradiance to plants and to the animals described by Sheldrake, as well as to objects. The organization of systems will always include a single director and the subordination of the individual to the greater prosperity of the whole.

Fields Ahead of Their Time

Curiously, among the greatest minds of humanity, the nineteenth century Italian theologian and philosopher from Rovereto, Antonio Rosmini Serbati, seemed to anticipate the idea of basic codes and informed fields in his *Doctrine of the Corporeal Principles.* Even though he did not know of quantum physics, he guessed at the existence of the basic code, with something he called the "corporeal principle." Rosmini felt the necessity to "discuss that what we perceive, some other virtue precedes" (the Renaissance neo-Platonists used this terminology), since the principle of the body is one "which is not present in perception but seems to hide itself behind the scene" (Psicol., n. 747). In fact, the physical body considered in itself, "not as we perceive it, could be an entity in itself" (Psicol., n. 775). He felt that there is something hidden behind the appearance of the body: "The *corporeal principle,* therefore, is not the corporeal substance spoken about by mankind when it uses the noun *body;* it is an unknown principle lying beyond this substance" (Psicol., n. 777).[27]

That something, Rosmini explicitly declares, is the very cause of the body that is regenerated by the body principle: "the corporeal principle is what determines the body" (Teos., vol. VI). It sounds like he could be talking about the basic code of things and its informed field: "corporal and material principle in a subject that is already informed by space" (Teos., vol. VI). Rosmini writes also about the nature of his code; he sounds like Pannaria when he talks about basic matter. Let us hear him:

> We are still in a state of illustrating the concept of pure principle. We distinguished the concept of pure principle from the concept of unidentified principle. The latter one is the true entity, but the pure principle is a very abstract concept . . . potential. . . . Since we think about an identified principle, that principle has become creature and is not a principle anymore, so is not what it was before: seeing that pure principle is pure entity, and pure entity is God. (Teos., vol. VII)[28]

That principle—which has *not* become a creature (which is no longer a pure principle)—is the field. Originally pure matter, when it is identified as a creature, it loses its purity. At the same time it acquires information and becomes informed matter, which in turn pushes the pure matter to combine. Intuition ahead of its time? Resonances? Precognition?

11 •—• EMOTIONAL FIELDS

Who has not had the experience of perceiving someone else's mood, the impression of feeling what someone is thinking or feeling, the certainty of knowing whether someone is lying or not? Who hasn't phoned somebody and been told that that person had been thinking about him or her right then? These are not coincidences. What about when we find ourselves saying "I knew this was going to happen" or "something inside was telling me"? Or the situation when someone far away seems to perceive the suffering of a loved one, or their fear or sense of danger? It is as if something inspires us, talks to us. These are examples of communication between another person's field and our own.

Emotions vibrate in the fields of organisms also, and they are responsible for the phenomena described earlier regarding plants and animals. A field can vary as a result of emotional stimuli and can also resonate with the emotions of other beings, because emotional states are also information. Just as an entity's informed field affects its shape, emotional variations produce physical reactions: laughing, crying, quickened heartbeat, excitement, rage, and much more. In the same way that the informed field communicates with the environment, so its emotional variations are transmitted to other beings. Our emotions affect the fields of our dog, our cat, our living room plant, the fruit on our table, the bacteria in our yogurt, the wine in the bottle, the bottle itself, the table, the walls, the air, and so on. Through the fields, we communicate emotions to everything around us: infinitesimal variations that the senses, generally, cannot even perceive.

Not just water but all matter has a "memory." All types of matter record events in their fields. Places can absorb emotional waves from dramatic events and retain memories, even across centuries. The frightening emotions that accompany a crime, for example, can become impregnated in the environment: house, plants, stones, earth, and so on. However, it is plausible that emotional communications mostly tend toward forms of life with which we already have an intense affective resonance (they are more probable with our own dog or plant than with the bacteria in yogurt).

Harold Saxton Burr observes that an acute suffering generates intense emotional waves, which readily influence the field of another being, while a slow agony, part of a more natural process, gets imprinted less intensely.[1] In general, nature tends to "fall asleep" in the presence of continuous rhythms, while it is awakened by sharp variations of rhythm, of tone, of intensity. This is also the same for the TFF: the frequencies of a medicine are imprinted with more efficacy if they are transferred in a pulsating way, with impulses and pauses. The cells respond better to discontinuous stimulations. Matter loves rhythm and is afraid of monotony; the fields endlessly palpitate. Certain rhythms render emotional transmission easier from one field to another.

In the case of an emotional connection or genetic link, emotional transmissions are not subject to spatial rules and may occur instantaneously, even at great distances, as we will see in the next chapter. Emotional transmissions are also not bound by time. Fields retain information, like recordings on a CD with unlimited capacity, or what Giordano Bruno defined as a "receptacle capable of accommodating itself to infinite substance."[2] An emotional event that was particularly intense can remain memorized for years and be "read" a long time afterward by a psychic, just as in the case of objects.

Emotion may even affect the activity of electronic equipment. An experiment with chicks published by René Peoc'h shows the influence of emotional fields on a mechanical device.[3] Let us see how. Chicks, by nature, demonstrate the imprinting phenomenon: they will follow the first moving object they encounter everywhere, as if it were their mother. In the experiment, a robot whose movements and rotating angles were chosen randomly by a number generator contained within the robot itself was placed in a cage with baby chicks. The chicks, emotionally linked to the robot as to a mother, tried to follow it. Peoc'h demonstrated that when the cage was empty, the robot moved in a random fashion, while when the chicks were in the cage with it, their presence influenced its movement (fig. 11.1).

The recorded impact of the emotional wave of the chicks on a random number generator demonstrates that the emotional field can influence nonorganic matter and objects. The world of "things" should be studied with new attention. The possibility that certain emotions influence matter can also be seen in the impact of intense emotions on the physical body (such as a heart attack), on the mind (sudden madness), and even on the environment (abrupt changes in the field felt at short

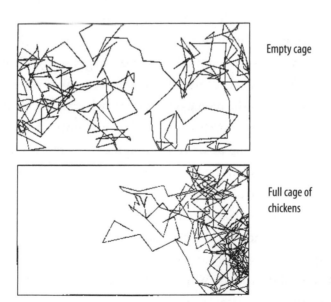

Empty cage

Full cage of chickens

Figure 11.1. Experiment by R. Peoc'h: the top diagram shows the robot's path taken in the absence of chickens; the lower diagram shows the same in the presence of the chickens. The emotional currents of the chicks affect the robot's movement.

and long distances). Some recent medical literature reports that intense and continuous emotions—experienced by some people who are a part of ongoing conflicts—seem to create irritations that in time may trigger diseases in the body.[4]

While certain emotions provoke disease, others cure. To cure, the emotion has to act spontaneously, beyond reason and will. It is the intention that makes the difference. The will to heal—the real one—comes not from the mind but from the unconscious, which belongs to the field. It is then that intense emotional currents can gradually destroy a tumor; if they do so in a very short time, we speak of miraculous healing. Maybe miracles do happen (but they are not reproducible) because in that situation faith and abandonment can emit intentional emotional currents so powerful as to disrupt, disintegrate, and then reform matter once again. These are field actions. If I imagine impulsively acting on one of my organs, nothing happens, but if I induce a deep relaxation, then visualize the organ until I "feel" that I am entering it to transform it, then the transformation of the imagined organ influences the real one. Isn't this similar to what was observed with Backster with the response of plants to human intentions? Not only real situations arouse emotions, but also thoughts, ideas, and images—any stimulation from within or from outside of us, any experience, real or otherwise.

Parmenides to the rescue once again: "what we think exists" (τό γάρ αὐτό νοεῖν ἐστίν τε καί εἶναι). Who has not noticed how the body is capable of responding to what we imagine to be true? The shy student, convinced that the girl he likes is looking at him, will find himself in a sweat with his heart beating rapidly, even if she is not there. Believing it is enough. If we are convinced we can be successful in an exam, we will do it; likewise, the opposite can happen. If we are convinced we are on the verge of a ledge, we can be paralyzed with fear, even if our feet are steady on solid ground. What makes the difference is the conviction, spontaneous and uncontrollable. Reason can convince the mind but not the unconscious, and even less the field. Both medicine and physics should study more the extraordinary and unpredictable human faculty that we call conviction.

The Importance of Conviction

There are more things in heaven and earth, Horatio, than are
dreamt of in your philosophy.

HAMLET

In these pages I would like to suggest that conviction is the key to the success of a phenomena that would otherwise remain unexplained: the *placebo*. A placebo is a medical treatment that has no specific drug action, but has a therapeutic effect through conviction. It can be used to treat a patient or as a control in pharmacologic experiments. The placebo effect is obtained if the patient is *truly* convinced that the medicine can cure him. If the medicine is appropriate the placebo increases the power; if it is not suitable or it is not even a medicine, there is an effect anyway. It is enough for the patient to be convinced and have faith in the doctor and the therapy; the expectation acts on the emotional field with unexplainable effects coming not from chemistry but from field actions. But conviction is something that comes from the soul, not from the mind. You cannot force yourself to have faith.

It is known that expectation can influence laboratory experiments: we have the tendency to see what we want, and it may happen that proof goes precisely in that direction. But what force are we talking about? Even the most controlled scientific experiments may always be influenced, though they are never repeated in exactly the same way. No precaution can exclude the placebo effect, exactly because of the relationship between conviction and the unconscious component of the field. If the scientist has faith in the result, even if he puts a psychological distance between himself and the desired result, he has a frequency that already may have acted without his knowledge, imprinting the equipment, the environment, and the event itself. Unconsciously we can influence any experiment, even at a distance, even if "blind," and the placebo effect may intrude even in psychokinesis. In an article in *Nature*, David Böhm pointed out:

The necessary conditions for the manifestation of paranormal phenomenon are the same considered optimal for an efficient scientific
investigation. Tension, fear, hostility inhibit the paranormal effect,
just as they hinder the scientific experiment in general and the prob-
abilities of success decrease considerably.[5]

World statistics on placebo medicines give percentages of responses
that can be compared with those of real medicines. Why not be curi-
ous (and if the researchers are not, then who should be?) and ask how
it is that a sugar tablet can cure like a real medicine? It is a shame that
this research is not in line with the economic objectives of the industry,
otherwise the fundamentalists of science would be forced to explore the
cognitive fields they fear so much. The placebo effect is a medicine, the
most interesting one.

Not only does the placebo depend on conviction, but so also does
its opposite, the *nocebo:* the conviction that a medicine is not suitable
can inhibit the effect, even when the medicine is appropriate. The faith
or nonfaith in the doctor and the therapy plays an important role in the
result one way or another, and our medical colleagues know this. This
is valid not only for patients, but also for those who give the medicine,
for placebo or nocebo effects also correspond to the faith the doctor
has in the medicine. Every doctor has experienced it. The *faith field*
changes the field to vibrate at healthy frequencies. If the faith fields of
the doctor and the patient are combined, they build a *success field* that is
more efficient than any placebo. The frequencies of this field then start
to influence the vibrations of the ill parts of the field, and they change
them, as in the scheme of the prevailing interaction. If the reciprocal
faith is sufficiently intense, the success field is a lot more powerful than
the medicine itself and can act as rapidly as a miraculous recovery.

The TFF is the first step to understanding phenomena such as
placebo effects and "psychic" healings. It is difficult to reproduce in
an experimental way the appropriate healing frequency, for it arises
spontaneously from a person's faith and desire to heal or to be successful at something. I repeat: if the precious frequency acts in an instant,

we cry "miracle." Maybe placebo and miracle are not so dissimilar.

A doctor reported that in order to heal an asthmatic patient, he ordered a new medicine from a pharmaceutical firm, with which he successfully treated the patient in question. Subsequently he gave the patient a placebo, but this had no result. Just when the doctor thought he had proved the efficacy of the medicine at the expense of the placebo, he was informed by the pharmaceutical firm that they had mistakenly sent him a placebo instead of the medicine the first time he ordered. It is obvious that in this case it was the placebo of the doctor that played a therapeutic role.[6]

Placebos are sometimes even used in surgery. In the 1950s someone tried to "pretend" surgical intervention by opening and sewing up the chest of a patient to treat angina pain. The result was that the person who had undergone the fake surgery experienced the same benefit as those people who had really been operated on. I remember one colleague who had to treat the healthy tooth of a patient who was complaining about pain. Only fake surgery stopped the pain of the neurotic patient, who was convinced of his pain and the efficacy of the surgery—in fact useless—that he insisted on.

The common rituals used to remove warts from the skin are numerous and often efficient, even if they are without scientific basis, and they have a strong following of convinced users. The ritual serves only to arouse faith in the healing. And if this works for warts, why should it not work for a tumor? Even the placebo tablet is itself a ritual to induce certainty, that precise emotional wave able to transform matter. It is an "honest deception."

I have experimented more than once with placebos coming from emotional waves. A neighbor, a patient of mine, asked me one evening to alleviate his unbearable sore throat, as he was experiencing almost a complete inability to speak. Not having anything useful at home, but knowing his faith in me, I thought about transmitting my intent to heal him into a bottle of water. A few hours after drinking it, my neighbor phoned me with his newly regained voice: the pain had gone. Another time, I was taking my daughter to Emergency. She was scared and in pain because of a small fracture. Not having anything on me, I sent the message to sedate and calm her to a bottle of water that I gave her. A few minutes later, her pain decreased and she calmed down, eventually falling asleep before we arrived at the hospital.

What happens with placebo? The conviction starts transformations at a cellular level that can, for example, stop an inflammation. Or you can inform water, modifying its structure, so that it carries the message. You may ask why there is a need to imprint water, if the intent is the active cause. The ritual is necessary to create conviction. The doctor *must* give something, water or tablet, even if he knows it is placebo, so he does not look like a quack to himself or to the patient. The gesture often starts the mechanism of conviction. In the Gospel we can read about simple rituals Jesus used when he worked miracles. For example, when he returned a blind man's sight he spat on the ground, made a plaster out of it, and spread it onto the man's eyes, saying, "Go and wash yourself in the pool of Siloam" (John 9:6–7). Jesus does not need any ritual; he does not need to be convinced. It is the blind man who needs to be convinced, and the show is for him.

If I am convinced that I will not burn my feet, I can walk on hot coals. If I am convinced that an iron bar resting on my neck and on that of my neighbor is butter, a simple coordinated effort of our two necks will be enough to bend it. Jesus himself, talking to the disciples who accompanied him, said, "He who believes in me will also do the things I do and will do even more" (John 14:12). The key is to believe intensely and absolutely. This allows the conviction to start and to become a force that does not ever recede, not even in light of evidence to the contrary. "If you believe and do not doubt," says Jesus. The Indian teacher Paramahamsa Yogananda wrote, "The world is only an objectified dream, and anything your powerful mind believes intensely occurs instantly."[7]

It seems that the waves of an emotional-intentional field can inform water, which can transfer them to anyone. This is done in some Indian ashrams where large quantities of water are activated during hours of group meditation; such water has some really extraordinary therapeutic powers.[8] Water "memorizes" information from emotional-intentional fields in a way that is similar to the TFF; water is also able to store and radiate infrared (IR) or UV or laser light or other electromagnetic waves.

The placebo mechanism is always the conviction—the *perfect wave* that can be generated in the doctor ("I believe in what I do") or in the patient ("I believe in what the doctor does" or "I believe that I will make it") or a combination of both. If this is missing, the effects can also be missing. More than once I have had the following experience: after a new treatment provided good results, my conviction of its efficacy pushed me to repeat the success with other patients; then, when my emotional level dropped, the successes were abruptly reduced. Afterward, the treatment continued to be less efficient; it seemed to become "burned out." When a series of successes with a medicine is followed by some failures, the placebo component of the conviction falls and inevitably collapses; from then on the medicine acts only according to its own characteristics.

A collective placebo effect has also been observed. When an anti-cancer chemotherapy was initially advertised as an almost miraculous medicine, success occurred in about 75 percent of the treated cases. But when the wave of enthusiasm became more habitual the success rate went down to 25 to 30 percent. When the emotional wave went down, the effect was reduced.[9]

Conviction also acts on the emotional field through *hypnosis;* this is a type of compulsory conviction, a "prevailing interaction" of messages being sent to a person through the hypnotic state. Among such literature it is worth mentioning a case in which the posthypnotic command to the subject in a trance was that his daughter would become invisible to him. This is exactly what happened: once he was "awake," the subject couldn't see his daughter sitting in front of him. She was so completely invisible to him that an object hidden behind her body was visible to him, to the extent that he could read the inscription on it.[10]

The senses are deceived during hypnosis. It is possible for a hypnotizer to transfer sensations such as tastes to the hypnotized subject: if the hypnotizer puts sugar or salt to her own lips, the hypnotized person immediately perceives sweetness or saltiness.[11] The hypnotized can sneeze if the hypnotizer smells ammonia, can feel pain if she gets pricked on the finger, blinks if a light flashes in the hypnotizer's eyes, and so on.[12] These are transfers of information just like in the TFF, interactions between two fields that at that time are blending with each other like those in the Kirlian image of the two leaves close together (see fig. 9.8).

We regard hypnosis as a type of prevailing interaction because virtual influences are accepted as real and they are capable of modifying matter as in other cases of field action. A hypnotized subject who is told that a pencil that touches his skin is a lighted cigarette will produce hours later at that precise point real wounds like those due to cigarette burns.[13] By canceling out the will and therefore the faculty of doubting, hypnosis allows the acceptance of any message as real. The same happens with small children who can be told, "Look, there is a gnome out there," and they will see it and be convinced about it. (Consider this: the entire world could be a virtual reality, and we could be as if hypnotized from birth, able to see and hear only what was allowed to us.)

If a diabetic under hypnosis subcutaneously injects himself with a phial of saline solution instead of the real medicine at the usual time, and says to himself convincingly that it is insulin, the blood sugar is reduced as per the typical action of the medicine.[14] The conviction acts in the same way as the medicine, which is only a molecular carrier. The true medicine is the information, which can act even with no molecules. It is a message in the form of signals assigned to the waves, which belongs to the physical world, not to chemistry.

Saints manifest stigmata wounds in their palms and not on their wrists, where in fact the wounds of a crucified person would have been (nails in the hands could not support the weight of a body hanging on a cross). But from the eighth century onward Jesus's wounds were depicted in paintings as being on his palms, so the collective imagination puts stigmata at the centers of the hands. This suggests not so much paranormal activity but rather information coming from human beings, carried in our collective consciousness. The stigmatic wounds that have been observed are almost always deep; they often open and bleed periodically. Sometimes they can be opened and closed on command, as in the report of an abbess from Umbria in the eighteenth century.[15] The mechanism of production would be similar to that which gives rise to cigarette burns in a hypnotic state. In both cases it has to do with images coming from the outside (as in hypnosis) or from inside (as in the stigmata), which are able to activate the information. The extraordinary intensity of the emotional field of a person manifesting the stigmata is able to recombine the structure of the skin, and the subcutaneous and muscular layers, to conform with the prefigured image. Drawings, images, numbers, and words can appear on a person's body, turning reality from imaginary to real.[16]

Some people have demonstrated the ability to live without food or water. Among many cases I will recall that of a German saint, Therese Neumann, who, in addition to having stigmata, did not eat or drink anything for thirty-five years. All this has been rigorously documented by an investigating committee sent by the bishop of Regensburg: the saint did not go to the bathroom, she did not dehydrate or lose weight, and she even spontaneously regenerated the quantities of blood lost due to her stigmata. She also regularly materialized the water and nutrients needed for life.[17] Such phenomena should be studied, rather than considered miracles or witchcraft, but in the end we prefer to ignore them, and research is once again defeated.

Some humans have psychokinetic abilities; in many cases they unconsciously move objects at a distance, elevate them, or destroy them. Their mere presence can break an apparatus, or cause explosions, fires, or other disasters. Such unconscious psychokinesis may result from the release of intense emotional currents that have accumulated in a very sensitive person.

It therefore appears that conviction, coupled with imagination and visualization, can transform matter through different manifestations; these are all traceable to the same phenomenon.

Therapies with Emotional Fields

The uninterrupted flow of becoming is a succession of images confined in time and space (otherwise they would be incomprehensible). We need references to distinguish the relationships between one thing and another, thus building a reality of classifications and postulates: minute, yard, zero, infinite. In this way life becomes a little theater that is represented by the boundaries of the mind: "here it starts, here it ends." If we could shift the terms, the angles, the poles, then new realties would emerge. Don't worry; there is no danger of ditching our world and our false certainties. We will carry on being trapped in categories in order to be able to continue to investigate, reflect, and communicate. My exposition here is just approximate and symbolic because some categories of matter are still missing: we can perceive the phenomenon of nonlocality, but we are not able to represent it. We must return to parables.

A field ends (in reality it does not end, we are just not perceiving it any longer) where another one starts, continuing the network of pure matter—the magician's hat, the horn of plenty that is never exhausted—where we can find anything. There are archetypes, ideas, and informed fields; there is what is expressed in our dimensions and what remains unexpressed. The net of the universe holds everything.

Consequently if you want to materialize something, it is enough to build an informed field. How? One way is to imagine it. This may seem simple, fantastic, and banal, but it is not easy to visualize in a correct way to generate a particular wave. If we want to dominate fields and matter, we have to act with images. *Visualization* is more concrete than naming; words can suggest images, but visualization has the power to produce emotions that can be transferred to the field and reach resonance. However, if you desire it too much, the healing emotion is not obtained (as the Chinese sage says, "Look for something and you will not find it; sit quietly and it will find you"). If you try to visualize when gripped by fear, you will fail because the fear of failure blocks success: one emotion neutralizes the other. If you try without conviction then you also will fail. Yet it is hard not to doubt or fear. "Do not fear, and do not doubt," says Jesus, identifying the keys to miracles of the emotional fields.

Faith, which causes miraculous healings, is an emotion so intense that it acts without any need of visualization, other than perhaps the actual one, instantaneous and real, of the one being healed. It does not come from the head but through something that everyone has tried at least once, which is very difficult to describe. *Therapeutic visualization* can change the nature of things. It can take months, years, or instants; it depends on the intensity of faith. In Sufism, imagination is seen as a kind of perception. Going beyond the limits of logic, we enter the analogical dimensions where nothing is impossible and where a simple mental image may have an impact on the senses similar to that of the thing itself.

A trance state enables a person to go outside the rules of the becoming and enter those of "somewhere else," thus becoming what doctor Alberto Sorti from Bergamo describes as "a mind spread throughout the universe that stores and memorizes each event (the forms, the frequencies encountered); it compares similar events and decides for the best."[18] For healing, Sorti started inducing a hypnotic-like state (cerebral alpha rhythm, the same as profound relaxation), which allows the unconscious of the patient to receive therapeutic messages. This is how they used to heal in the Schools of Hippocrates in ancient Greece: once drowsiness was achieved, induced by hypnotic herbal infusions, the doctors talked to the patients during sleep, sending reassuring, persuasive, and healing messages; the ill person then had the feeling of having dreamed those messages. Sometimes the messages were put into the mouth of divinities to exploit the authority of the Verb. Psychologists speak of the super-ego, but in fact psychology has roots much older than the nineteenth century!

A patient can learn relaxation and visualization techniques and practice sending information to his or her own body. The secret of the miraculous phenomena is not in the thought or in the word, but in the images that evoke emotions. The word alone will never be as efficient as visualization. The patient also needs to have a real desire to heal and to live, which is not easy since he has fallen ill precisely because his own emotional field had started a program of self-destruction. Unconsciously the person does not want to be well, but instead wants to die, even though he states the contrary. This is the sad reality of the majority of cancer patients. Finding the will to live again is not done by reasoning with the mind, but with the heart, the irrational, the field. Sometimes it can be perceived even from the outside, this new, powerful, upsetting, incontestable force. It is absolute conviction.

I witnessed the healing of an advanced tumor using this technique. It was a case of a second relapse of colon cancer. Mirella, forty-three years old, had only been partially operated on. She had also been treated with third-line chemotherapy and radiotherapy. Her life prognosis was about six months. At that point, she decided to transform her life, to suffer no more of the emotional conflicts that had allowed the disease to develop. She rebelled against her own emotional slavery; it was a war of true liberation. She started a program of relaxation with visualization twice a day, for at least an hour, in which she imagined hitting the tumor cells with light rays to destroy the cancer. It was like using radiotherapy in her imagination. The results were extraordinary and clearly evident in subsequent radiographic tests: the tumor mass was progressively reduced and became more placid. In a few months it disappeared. Checkups in the two following years confirmed the absence of the disease.[19] Three years later, she was back in her previous emotional conflict. She tried using visualizations again, but she couldn't do it anymore (the renewed program of death this time did not change); she underwent another, and in this case useless, chemotherapy session, and in less than a year she died.

It may seem nonscientific to cure tumors by therapeutic visualization, but it has its own logic, as well as its difficulties; when the conviction is lacking, as in Mirella's case the second time around, it will not work. This is an isolated case, with no statistical value, but it is still *a* case and deserves the weight of serious study. Her emotional fields acted as an "imaginary" radiotherapy, using the low-frequency fields. Similarly, a team of Italian researchers led by the physicist Santi Tofani and the engineer Fausto Lanfranco obtained significant reductions in tumor mass, in vitro and in mice, which stabilized over time, through the use of magnetic fields.[20] Whether generated by visualization or by a device, there are frequencies that can assist chemistry in medical therapies, but a paradigm shift in scientific thinking is needed to take advantage of them. Doing so is called for based on information that presently exists.

Sixth Sense

We have seen that sensations allow us to distinguish the solid from the empty and that perception is partial and subjective. The senses are the ones that determine where the solid ends. However, are we sure that emptiness starts at the place where the senses cannot "feel" anything? What are the real boundaries of bodies? If we observe any surface under a microscope, we cannot establish its limits exactly. The senses are not the criterion for scientific and certain evaluation.

Bodies extend beyond the boundaries of sensitivity, but in the form of fields. How far the fields extend is difficult to say. Furthermore, the extension constantly changes because the bodies live, the molecules dance, and the fields pulsate. How can the fields be measured? How can we establish the boundaries of things? When does a sound end? When does the ear not hear anymore? When is any instrument not capable of registering anymore? Why should the death of a sound coincide with the limit of an instrument? The sound tends to zero, but it is not cancelled. The scents, the smells, are they exhausted when we do not sense them anymore? And the things we see? The invisible is not empty, but is a crowded crossroad of information exchanges and fields of all kinds; some are expressed in physical bodies, while others exist as "pure form with no substance." The universe extends beyond the bounds of our instruments of measurement; the world is certainly vaster than what our senses perceive.

The field can manifest itself in various ways. We have all felt an impulse to turn around and look because, without having seen him or her, we feel the gaze of another's eyes upon us. We say it is because of a "sixth sense," but this does not really explain it. The term *sixth sense* is just an expression for the field, for *something* that talks to us from far away. It may warn us of danger, and it often precedes our senses. We can see it functioning for all cats, dogs, and children; if we stare intensely at a dog he will move toward us. If I forget the roast meat in the oven and do something else, suddenly I *feel* as if I have to run to the kitchen to save lunch at the last minute; it was my field that resonated.

Thinking of these occurrences as resonating fields may seem strange because we are not yet accustomed to thinking of condensed matter as a concert of frequencies. So instead we call this field that resonates with everything that concerns us instinct, or sensitivity. Still, every time that we "have a sensation," it is the field that has alerted us. True knowledge is much greater than what appears at the threshold of consciousness. Some call it unconscious, but it is an expression of the field.

Psychophysical balance is important for good-quality field transmissions. When the transmission mechanisms are maintained in good order, the information from the field registers in consciousness. Otherwise, it cannot cross the threshold. Although we possess a field that is continuously informed about what concerns it, even from a distance, it is difficult for us to realize this. Over time we have lost the faculty of feeling through the field instead of the senses. Only when the water of a lake is still, not when it is choppy or stirred, can we read what is written at the bottom.

The Q'eros, for example, perceive more than we do because they live far from psychological and technological stress. When they divine with coca leaves, even before they throw them on the mesa, they know what they are about to read: their field has already resonated. I have personally heard them more than once anticipating the response of the leaves. Resonating fields are at play whether a person is reading tarot cards, coffee grounds, oil left dripping on water, or water itself, or using the I Ching, a planchette, or a crystal ball, or any of the means possible and impossible used throughout human history. How many times, without noticing it, we read the field of others! When two people feel attracted or repelled, it is their fields. Sympathy, antipathy, feeling, the sensation of being able to understand, or that of having a "wall" between us and another—these are all field resonances.

12 •— THE GREAT MOTHER'S CLOAK OF OBSCURITY

I just closed my eyes and there rose from the middle of the sea a face divine that lifted toward me a countenance that even the gods must adore. Then little by little I seemed to see a shining image emerge from the sea and stand before me. . . . Abundant long and wavy hair streamed gently down . . . the divine neck. The top of the head was bound with a crown of interlaced wreaths and varied flowers. In the midst of it, just over the brow, shone white and glowing a round disk like a mirror or a miniature moon. In one of her hands she bore serpents and in the other ears of corn. The vest had many colors, now gleaming with a snowy brightness, now yellow with hue of saffron, now flaming as a red rose. But it was the cloak cast around the body that dazzled my gaze far beyond all else, for it was of deep black, which shown with a dark glow. . . . On the embroidered hem and on its surface also were scattered sparkling stars, and in their midst the full moon glowing with fire . . .

APULEIUS OF MADAURA,
THE GOLDEN ASS

This is how Apuleius of Madaura described pure matter in the third century B.C.E. Lucio, protagonist of *The Golden Ass,* recounts a dream in which he sees Isis, Mother Goddess of all things and nurse of living beings.[1] She has the attributes of all Great Mothers: moon and snakes, nurturing corn, red roses, and the divine cloak embroidered with stars. All these are found also in the Celtic goddess myths of Avalon and in the Christian Madonna myths. The cape is dazzling not so much because of the stars, but because it shines with a dark glow. This oxymoron has the flavor of initiation: the divine Mother is black, *luminously* black.* No man can bear the sight, because it is *black light,* the inner energy of matter and the secret light that the alchemists hoped to find. Black because it is invisible and the true color of light is black. Cosmopolitan, an eighteenth-century alchemist, wrote that in reality the sun is a cold star and its rays are black.

The cloak of the Goddess is the obscure matter that the senses do not perceive, a sea of energy that did not originate from the big bang and does not vanish even if a small part is combined in mass. It is not heavy matter, "subject to gravitational laws different from those known."[2] The black light is pure and powerful because in its dark glow hide the essences of all things: information, codes. It is the Goddess's *luminous* dark cloak.

Pure Matter and the Exchange Principle

Do you remember how our journey started? There are two states of matter: one that combines into bodies and is perceivable, definable, measurable, and constantly changing (mass); and the other, indeterminate, *pure,* uncombined, which does not enter into the becoming; it is invisible and cannot be measured or imagined. Pure matter is a continuum, the matrix of all masses, the invisible connective tissue between them. Just as in the microscopic world the fields allow the exchange of information between bodies, in the macroscopic one, the omnipresent

---

Splendescenes atronitore, "Glowing in black light"

tissue of pure matter renders every communication possible.

Let us see how. You remember that pure matter is free of time (because it is lacking mass) and of space (because it fills it completely); it is the continuum of Parmenides, omnipresent and eternal. By extending to all areas, from the atomic to the sidereal, pure matter is a network that keeps the universe in order and makes it exist. "Being created to sustain the bodies," Giordano Bruno wrote about single and continuous space. Pure matter allows information exchange by connecting all the parts and being everywhere simultaneously. Every part communicates with everything and everything with every part. The network has no limits of capacity and can store and transmit infinite information simultaneously and never become saturated. Bruno wrote:

> The substratum of these principles is a unique infinite space, capable of storing in itself infinite substance, in which something can exist. In the same way these visible entities, that come to touch our senses, entirely fill this space, the extent of which is equal to the size of the bodies, that new space, which proceeds beyond the limits of this space and extends to infinity, is certainly endowed with a receptive power not less than the other, and nobody, if he is not completely ignorant, will question the truth of this statement.[3]

The areas where the two dimensions (space-time, and the non-spatial and nontemporal) interact are the fields. Pannaria's *principle* of *exchange* establishes that every physical interaction has a mediator field, the space between bodies, through which matter communicates with other matter, and that living systems are continuously crisscrossed by matter and energy flows: "A continuous flux of exchanged particles is already a field, which means that the particles that are generated by a body direct themselves toward one another, the presence of each influencing the other."[4]

Even when it seems to occur by contact, the exchange of matter and information is always at a distance, small or large; it takes place in the so-called *zone of the middle,* the space that appears empty. According to Pannaria, the field is a continuous occurrence of exchange, continuously generating particles. Bodies always exchange "between the two"; the state of exchange is in the "between," in the middle. The exchange is then evident because something extinguishes on one side and begins again on the other side, proportionately.

Pannaria believed that the field belongs to the physical world of pure matter, and he noted in 1992, shortly before his death, "It follows that the vacuum is pure matter. The small and infinitesimal voids are pure matter, home of small and infinitesimal exchanges between elements and sub-nuclear particles." The bodies are immersed in pure matter that, keeping the parts united, allows communication so that everything takes part in everything. The universe is an exchange that moves, generates, renews, and reproduces. The informed field is "that third kind" (which Plato spoke of), which connects the two aspects of the physical world: the scene (combined matter) and background (pure matter).

Plato in the Timaeus distinguished between being, which is immobile, eternal, and intelligible by pure thought (pure matter), and becoming, which is changing, subject to time, and the object of sense perception (combined matter); he defined the world as "a living being, with a soul and intelligence."[5] Then he tackled the issue of generation and creation. Between the two dimensions (continuous and fragmented), Plato imagined a space where the bodies are generated: "it admits of no distinctions and offers a place to however many things are generated;" it is not perceivable but "hardly an object of faith."[6] The Demiurge (he who made the world) mixed the two essences (the indivisible and the divisible) to form "a *third kind* of intermediate essence and constituted it in *the middle of them,* between the indivisible and the divisible in the bodies . . . *and it filled the intervals*" (sounds like Parmenides). He called this *third kind* "a receptive nurse of all that is generated." It is the mother of all created things, the mold that gives shape.

Therefore we have to recognize what is generated, what it is generated from, and that which is created in imitation of what is generated. "What is generated" is combined matter; "what it is generated from" is pure matter; "that which is created in imitation of what is generated" is *informed matter* from which the basic codes are made, which in turn inform the fields. The informed field acts as a bridge between the two worlds. It proceeds from the continuum and, selects the information to combine the right elements together to produce the mass. The field is informed matter that comes on the scene thanks to the body. Organized and directed from the informed field by the code, mass in turn maintains the field, exchanging with it and serving as a support in the space-time dimension. Exchanging information, field and mass manage to maintain their identity and physical integrity, and the information can radiate into the environment, communicating between fields.

Harold Saxton Burr already discovered that certain diseases affect the energetic field, and this was interpreted as the effect of the disease itself.[7] What if the contrary is also valid? What if the mutation occurred in the field before it was transmitted to the body? What if the basic code decided to get ill and then heal itself? The code drives everything; in the role of the designer, as a hologram, it comprehends more than what appears. Which process otherwise would allow the salamander to re-grow its tail? Or heal a heart that has failed? Or teach stem cells into what cells to differentiate? Although the mechanistic approach is not able to explain it yet, it is the field that recreates the shape that has been disturbed or establishes how a cell must differentiate.

Things Communicate at a Distance

It is plausible that things do not end at the point where the senses cease to perceive them. Even the earth is submerged in an electromagnetic field (the Schumann waves and the Curry and Hartmann networks are expressions of this), which reaches far from our planet for thousands of miles: to where? Gravitational fields seem even more extended, governing the orbits of satellites around the planets and the planets around the sun, even at great distances (think about Pluto). What extraordinary strength the gravitational field has! Where does it end? Constellations millions of light years away, dim lights in the night sky, how can they exercise the influence here described by astrological science? What forces are we talking about? There is still something in the concept of the field.

Imagine two particles coming from a common source, such as two photons emitted by the same atom, being abandoned to different destinies, distant enough from one another so that they cannot interfere with each other. We then discover that the two still maintain a relationship as if they were in contact, so much so that what happens to one is transmitted immediately to the other: if force is applied to one of them, the other reacts simultaneously. This is the Einstein-Podolsky-Rosen (EPR) paradox, which was published in 1935. This was Einstein's attack against Bohr's *Copenhagen interpretation* of quantum mechanics, in which he had, eight years earlier, argued that the world, in order to exist, must be observed.[8] For Einstein, existence was independent from he who observes it. He felt that there was something fundamentally incorrect with quantum mechanics since it predicted violations of locality. In other words, Einstein denied that an event could produce instantaneous effects at great distances.

Instead the opposite happened, as they affirmed both the quantum theory as well as the principle of nonlocality. Thirty years later, in fact, John Bell, a theoretical physicist at CERN in Geneva, produced a mathematical formula, not yet accepted by all physicists, that demonstrated the possibility of instantaneous interaction at a distance with the character of nonlocality (*Bell's inequality*). The *principle of nonlocality* seems to be confirmed by the experiments on plants and animals by Backster and Sheldrake. Certain organisms know how to stay in touch with others of their own kind or social group, even at great distances; if this principle also applies to elementary particles, it is plausible that nonlocality is a universal law.

To support the theory of nonlocal interaction between particles, in 1982 the French physicist Alain Aspect of the Institute of Optics at the University of Paris conducted an experiment. He heated calcium atoms with a laser, which produced identical twin photons, which were made to travel in opposite directions toward polarity analyzers. In accordance with quantum theory, Aspect discovered that the polarization angle of each photon was always correlated with the other. This means that either the two photons communicate faster than the speed of light or that their connection is not local. The Aspect phenomenon cannot be explained within the traditional scientific paradigms. The same is presently true for all phenomena relating to nonlocality.

If there are no electromagnetic currents to connect the photons, then these phenomena must be due to field resonances. The field theory can justify phenomena of nonlocality in animals (humans included), in plants, and even in elementary particles. There are still several situations in nature that can be explained only by nonlocality: magic, dowsing, radionics, and paranormal phenomena. In nonlocal communication, for example, we can recognize principles similar to those that underpin sympathetic magic: among the postulates is the idea that things can act together at a distance, by resonance, and that transmission occurs by means of "what we can conceive as a sort of invisible ether . . . through a space that appears to be empty," as Sir James Frazer writes in his historical text on magic and religion. He adds, "Things which *have once been in contact* with each other *continue to act* on each other at a distance *after* the *physical contact has been* severed."[9]

This law of contact governs what is called "sympathetic magic." An example is the relationship between a person and body parts that have been separated from him or her (hair, nails, teeth, placenta, umbilical cord, and so on). Think of the custom of saving children's teeth for an imaginary fairy to take away (to allow the growing of much stronger ones). These magical relationships also occur with external things, like tools, weapons, manuscripts, prints, and objects of various kinds. These are interactions between the animate and the inanimate worlds, and they are not explainable with classical deterministic principles.

This is not the right place to refer to the universe as "magical"; I would prefer to call it "unknown" because magic is nothing but natural phenomena that for now are neither demonstrable within the predominant scientific paradigm, nor reproducible, nor statistically significant, but they exist. When we encounter unexplainable realities, we cannot ignore them or dismiss them as magic. What does *magic* mean? It means that we are still not yet able to explain a phenomenon that should be studied.

Something continuously brings together the extraordinary events regarding organisms and things in general. Even the earth, as Lovelock says, has its forms of expression. As do atomic particles, the molecules of medicines, cells, and all forms of life. It is always the field that acts in both the short and long range, but to what extent? Where does the effect of a star end? Where does the paradox of the particles end?

One explanation is in "quantum nonlocalization." If short-range cellular communication (biophotons, ultraviolet radiation, mytogenetic waves, and much more) is an electromagnetic phenomenon, long-range interactions (between cells and organisms, plants and animals, plants and humans, animals and animals, animals and humans, humans and humans, and between living beings and things) can be explained by assuming a different type of force, which Sheldrake called *morphic resonance.* The term is derived from *morphé,* "shape" in the Greek language, because the morphic field—just like the informed field—controls the shape. We have seen that despite the removal of a part of a leaf, when the leaf is photographed with the Kirlian method, the luminous halo manifests itself around the outline of the whole leaf, which confirms that the field consists of morphogenetic and structural information. Whether or not there is combined matter, things exist by virtue of the basic codes.

For Sheldrake, morphic fields coordinate and confine the various parts of a system in time and space, preserving the memory of past events. Resonance is the way in which memories are transferred through space and time. The morphic field has elastic characteristics: it tends not to break even when the individual is far from the places or beings with whom it has an emotional resonance or genetic affinity (fig. 12.1). In this way, a connection is maintained, unconscious contact between the individual and everything within its field. This is the only way to explain the exchanges of emotions and information, and it could also explain paranormal phenomenon.

Sheldrake points out that if an animal belonging to a group finds some scattered food, awareness of its location reaches every individual in

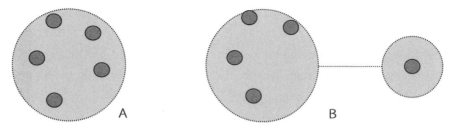

Figure 12.1. Scheme of a morphic field of a group of beings (A)
that remains connected even as one is moving away (B).

the group by traveling through the morphic field. Just like the particles of Alain Aspect. The same is true with fear reactions: if a member of an animal group is threatened by danger, even at a distance, immediately all the others manifest apprehension. The individuals are connected by their fields to their places, their things, their loved ones. Even when they are far away, their fields foster their emotional connection; through them they *feel* something happening in those places or to those people.

Regarding nonlocal communications, besides the elasticity of the field supported by Sheldrake (which still has a spatial component), we can think of the continuum of pure matter as having no characteristics of locality; rather it is "everything everywhere simultaneously." It is an ideal network for instantaneous communication in any part of the material universe: it is sufficient to enter the circuit, and the vibration resonates with others. There is no distance in the great sea of pure matter: fields have no boundaries in the continuum of pure matter, the black cloak of the Goddess.

The fields therefore can resonate with intentions, record and memorize events, enable communications at long distances with characteristics of nonlocality, and remain active in the state of waking and sleep. The degree of resonance is proportional to the intensity of relationship. Think how twins can *feel* each other at a distance, or mother and son, or two people in love. Resonances can also occur suddenly and unexpectedly, with unusual combinations, resulting in apparently random precognitions. Countless natural correspondences, rhythms, and impulses drive the course of events: they are just the tip of the iceberg.

What we perceive is only nonbeing. Being is barred to our senses, so we have to make do with shadows. Maybe some rhythmical sounds, percussion and so on, are able to connect to the rhythms of the fields. In magical rituals we find dancing, music, singing, rhymes, repetitions, and other rhythmical expressions. Even Ighina argued that life is pulsation and rhythms, and by using certain rhythms we can transmute matter: music can have a devastating or repairing effect as numeric sequences.

Experiments on migratory birds and monarch butterflies suggest that memory kept in the fields continues to resonate with ancestors from previous generations. But new patterns can also be formed. Remember the example of forced deviation in the migrating birds' path in which the next generation followed the new path, having developed a new memory to replace the previous one? Similarly, in the world of pills and metals, a new message can cancel out the memory imprinted by the TFF. If birds can establish a new migratory cycle in a single generation, this does not imply genetic mutation but rather the functioning of the field memory. Think about the so-called collective memory of our ancestors: entire groups—families or tribes—inherited knowledge unconsciously that did not seem to extinguish either over time or in space, and has been transmitted to succeeding generations of individuals. This type of memory allows the individual to tap into the wealth of memories of a species and to contribute to it in turn. Many rites, ancient or in use, might be able to be traced back to this. I will only mention Jung's theory of the collective unconscious, while leaving unexplored Frazer's writing on the tribal rites of the ancestors.

Natural Resonances

If it is the basic code, through the field, that designs the shape, combines the matter, and organizes the system, and if it really safeguards identity and balance, and enables all parts to communicate with everything and anything within the environment, and also in nonlocal situations, then it is reasonable to think that mass is only the last event in the process via which a code appears in the space-time dimension.

Nonlocality is made plausible by the principles of quantum physics: two parts of a system, separated in space, stay united through a quantum field of which we cannot calculate the exact spatial extension. This would explain the phenomenon of fields examined so far and even very ancient sciences such as *dowsing* and *radionics,* which are based on the principle of resonance.

Dowsing (radioesthesia) owes its name to the Latin *radius* (ray) and to the Greek *aesthesis* (sensitivity). It teaches us how to perceive through natural resonance the radiation emitted from bodies. It allows, for example, the discovering of veins of water or other substances underground (minerals, oil, and so on) by use of a "rod." The man, who, equipped with a wooden stick shaped like a rudimental slingshot, goes around the countryside in search of underground water—a scene that has been repeated for thousands of years—is not crazy or deluded: he is applying the physics of the resonant fields.

Even radionics—which allows diagnosis and cure at a distance—is based on the same physical principles. The apparatus used works without electricity, employing a so-called witness (such as a photograph, a sample of hair, blood, or a similar thing). Radionics has an ancient history, but only in the last two centuries have researchers like Abrams, Reich, Lakhovsky, Delawarr, and others reproposed to study it.

In the early years of the twentieth century, the French researcher Georges Lakhovsky went back to the theory of the Nobel Prize winner Albert Szent-Gyorgyi on the bioelectronic nature of the organism, which considers the cells as oscillating circuits capable of resonating with electromagnetic waves. Oscillating circuits are what allow a radio or television to function. Given its tiny size, a cell is able to emit and absorb radiation at a very high frequency, so (according to Lakhovsky) all organisms thus exchange radiation with the environment. The electrical phenomenon of oscillation is explained once again by the laws of resonance. The oscillating circuits of Lakhovsky are circular coils of electric wires and condensers that, vibrating at a given frequency, produce an electromagnetic field that can resonate with diseased cells by reordering the oscillations. Some of Lakhovsky's devices can accelerate cell division in plants and reabsorb excessive growth due to parasites, while others delay germination and slow down the development of the buds.

In Lakhovsky's view cancer could be caused by unbalanced oscillation of the cell. Some plants that became infected with cancer after being inoculated with *B. tumefaciens* were then subjected by Lakhovsky to the action of oscillation circuits; as a result, their tumors became progressively necrotic and dry.[10] He has documented cases of regressed tumors where special oscillating circuits have been connected to the bodies of ill people.[11]

Georges Lakhovsky considers the atom an "energy vortex" (a little bit like bubbles in fizzy water) and matter as an ensemble of electromagnetic radiations of different frequencies on a "vast keyboard."[12] He writes about a primordial substance (which he calls *Universion*), penetrating between the molecules and the particles of the bodies, forcing them to remain at certain distances from each other and to rotate at predefined speeds.[13] According to him this primordial substance is identical everywhere (therefore homogeneous and continuous, just like the pure matter of Severi and Pannaria) and may operate as an ordering field for matter itself. Even for Lakhovsky, the world does not exist in emptiness, but rather as a continuum from the interplanetary spaces to the intermolecular ones, filled with energy.[14] Just as water is made of denser parts and more liquid parts, so combined matter is an alternation of solids and empties, masses and fields.

For Lakhovsky the atom is an "oscillating circuit," and it is the energy in the surrounding emptiness (the field, we would say) that makes it oscillate. This reservoir of all energies also continually generates and energizes particles. Wilhelm Reich is also convinced that "subtle energies" are everywhere, permeating both bodies and emptiness. He writes that everything is alive, including magnets, crystals, light, and heat. Together with Reichenbach, he thinks that these subtle energies are spread from the field as radiating energy.

Dozens of books would not be enough to contain the experiences of years of research into radionics from around the world. There are still many scholars of this science, from university graduates to beginners, always careful not to be found out, because the witch-hunt is still on. One of the principles of radionics is that the part vibrates with the whole or with other parts, regardless of distance. There is interesting literature on the experiences of people hearing or seeing things, people, or events thousands of miles away and even sometimes interacting with them.

We have already looked at the phenomena of extrasensory communication—such as in the sperm that reacts to the suffering of its distant donor—which illustrates the principle of the "part" that vibrates with its "everything." The contrary is also valid: the suffering of a part extends to the whole. I refer to treatment (good or bad) at a distance through contact with even a minimal part of the body (nails, hair, blood, or other). By contacting the witness of a person (a photograph, part of the body, or clothing or something that belonged to the person), it is possible, under certain conditions, to investigate his physical and emotional health, even from a great distance. The witness is necessary to establish contact with the investigated person and to *read* his field. Pendulums, circuits, and forms of any kind are all tools to better identify the resonance between the field of the researcher and the subject.

Traces of a person's field have also been found on things owned by or that came in contact with the investigated person, just like fingerprints. Many people know what it means to feel what an object emits that has been in contact with a patient in agony up until the moment of death: waves of suffering are "registered" up to that mysterious final event.

Everything leaves traces of itself on objects, in the field, on people, and in places. In the print there is the code. In practice, it is possible to know a thing's basic code even in its absence, just as we can send the basic code of a medicine to a person even without the medicine. If we exclude charlatans and swindlers (as in everything), this is not magic but physics in its infancy: the imprinting of a person's field on objects, clothes, and whatever else he or she contacts takes place according to the interaction model between informed fields.

More than anything else, photographs are imprinted with the frequencies of a subject, through which a radionics expert is able to enter the basic code of the portrayed person. The information imprinted in the image is not limited to the moment of the photograph, but it includes also information of past events, and even future ones. Radionic readings have been tested in which information on events that had not yet happened was obtained from a photograph taken years before. This indicates that the witness is only needed to establish a connection with the other person.

Information can travel the other way as well, in which the witness is instead used as a transmitter for thoughts, messages, therapeutic frequencies, medicines, and so one, to any part of the world, even without the subject knowing it. As in everything, the other side of the coin is the possibility of sending harmful frequencies and causing pain. This is why some people bury nails or hair, so that nobody can use them for magical rituals against them. The magical operations—as Giordano Bruno said—have to be carried out in the empty spaces, where the fields vibrate. Knowing how to operate there, we could scientifically create "filters of love" or induce therapies at a distance or, on the contrary, cause suffering, disease, or death. Black magic is nothing but the other side of good magic, while magic in general is the other side of physics, physics described with the enchantment of a child or of a poet.

Radionics can transmit information, because in reality there are no distances. In fact, what is taking place is not even a transmission. If the distance depends on space and time and these dimensions are relative and illusory, then the movement across them is also illusion. It is nonbeing, while reality is found in being, that famous continuum, lacking space and time. Even if distances, movements, and transmissions are only conventions, we will carry on talking about nonbeing as if it were real, because there is no other way for us to express that we inhabit this nonbeing. We shouldn't be amazed anymore about the phenomena of nonlocality and nontemporality; we should instead study them.

We should also reconsider the significance of symbols and archetypes. According to Wolfgang Pauli, a quantum physicist and Nobel Prize winner:

> The findings of modern physics have accordingly led to a fundamental change in the attitude of modern man toward the archetypal ideas that are at the basis of matter and energy. Since time immemorial, the idea of matter had been closely linked with the Mother archetype. In alchemy, there was an elevating of the status of this idea in that the prima materia was actually assigned the attribute of the Increatum, which orthodox Christianity had assigned exclusively to God, as the masculine spiritual principle. When new physics ... demonstrated that what had earlier been known as "material substance" was in fact ephemeral, materialism was deprived of its very foundation. This "substance" has been replaced by the law of conservation of energy by means of which mass and energy are recognized as proportional and hence equivalent (inertia of energy). . . . In this connection, it seems significant that according to quantum physics the indestructibility of energy on the one hand—which expresses its timeless existence—and the appearance of energy in space and time on the other hand correspond to two contradictory and complementary aspects of reality.[15]

We are once again at the world stage and backstage of Pannaria.

A symbol evokes a different idea than an immediate, perceptible thing; it has a representative function. Kant understood it as an "intuitive and analogical representation"; for Goethe it is an image in which the universal is grasped in the individual, which embodies it inseparably, in a game of mutual referrals; for Jung the psyche expresses the most profound contents of the unconscious (we would say the field) in symbolic form. Certainly, the archetypal images and the symbols produce the same effects on the field of the soul of the *thing* they represent: they are messengers of information.

A symbol vibrates at the same frequencies as that which it represents, so (whether it is a number, sign, or other form) it talks directly to the unconscious: it appears, resonates, is understood and answered. The unconscious (or field) answers the symbol by modifying its own vibration, detectable in different ways, such as involuntary movements of the arm (if the subject holds in one hand an amplifier, such as a pendulum, for example, a variation of oscillation will signal the obtained resonance). Or it might be detected with kinesiology techniques, auricular medicine, electro-acupuncture. In all cases the same resonance that takes place with physical objects happens with symbols written, drawn, or even just thought. Thought is a symbol that resonates with the field the same as writing: thinking or writing awaken the archetypal power to act at a distance (which does not really exist). The field accepts the symbols and can thus look beyond matter and space.

Importing the Effects, Not the Things

According to radionics, every entity emits waves, called *waveforms,* with frequencies, intensity, and shapes of their own. These very weak perturbations of the field are generated by the shape of the thing, especially those that are oriented along the geomagnetic axis. Thus, the objects, geometrical shapes, drawings, and buildings (from the Egyptian pyramids to the Gothic cathedrals) can induce negative or positive biological effects. Crystals vibrate and emit phonons (quanta of sounds) that we do not hear. DNA and molecules generally emit sounds we do not perceive. The earth and the stars pulsate in the "music of the spheres" that escapes the human ear also. All matter is made of very fast vibrating waves that escape the senses. Matter is "the crest of a wave, curling like the sea."

The most ancient alchemistic and philosophical tradition maintains that bodies are immersed in a fluid called *ether,* an immense connective tissue like a computer network, through which all things are connected together. Every body is immersed in the ether, in which it creates a *minus* of equal shape and dimension, like a mold, such as when a body is immersed in water. And just like in water, waves propagate in the ether. Full of information, they spread in the network, and so everything remains informed about everything, and everything communicates with everything. Just by being there, every body is a source of perturbations that are transmitted through the connective tissue of the universe, which other bodies can "feel" at great distances.

The Italian physicist Luigi Borello envisioned, as a mechanism of transmission through the ether, a progressive deformation of packets of neutrinos, which make up the ether.[16] This is a thought that was derived from the neutrino theory of Cesare Colangeli (1950), according to whom the field, the electromagnetic waves, and matter are just polarizations of the elementary particles: mobile (as in a series, one after the other) in the case of electromagnetic waves, forming static nodules in the case of matter. Borello points out that "there can be no vacuum," since "empty" space in a quiet state, electrically neutral, consisting of two opposing electrical charges (which therefore have no action) would form a neutrino.[17] The neutrino is a particle with no mass, consisting of two heteronymic, that is opposite, charges. Just like the pure matter of Severi and Pannaria, the ether would be a continuum of neutrinos where transmissions do not happen with movement but rather progressive polarizations disrupting the continuum.

Let us imagine this elastic and deformable fluid in which a body moves a fluid mass equal to its volume and shape. The surrounding fluid bends in a centrifugal effect, tipping over neutrinos, which tip over their neighboring neutrinos, like dominos in every direction. A shape always reproduces its own shape in the universe, especially when it is in motion. This wave propagation is not a movement; it is like the "ola" that seems to be a moving wave though its elements are not moving.*

According to Borello, Colangeli realized Einstein's dream to find a formula to define matter and field together. Einstein wondered about

---

*The "ola" is an example of the metachronal rhythm achieved in a packed stadium when successive groups of spectators briefly stand and raise their arms. Each spectator is required to rise at the same time as those straight in front and behind, and slightly after the person immediately to either the right (for a clockwise wave) or the left (for a counterclockwise wave). Immediately upon stretching to full height, the spectator returns to the usual seated position. The result is a "wave" of standing spectators that travels through the crowd, even though individual spectators never move away from their seats.

the physical criteria that distinguish matter from the field and came to the conclusion that a qualitative distinction between them was impossible. This is what he wrote:

> It makes no sense to attribute different qualities to matter and field. We cannot imagine a surface that distinctively separates field and matter. . . . In our new physics there would be no more room for the pair of field and matter; there would only be one reality: the field. This new view is suggested by the great achievements in the field of physics, as well as the successes registered in the formulation of laws of electricity, magnetism, and gravitation, in the form of structural laws, and with the recognition of the equivalence between mass and energy.[18]

Drawings, pictures, and words emit waves similar to those produced by the thing they designate. Symbols are examples of how a sign can evoke universal archetypal functions. Celtic runes, Chinese hexagrams, Hebrew letters and numbers, Tibetan mandalas, tarot Arcana, alchemist symbols—these are all signs that emit powerful waveforms and can replace the archetype they represent. The Italian physician Pierfrancesco Maria Rovere wrote, "To capture these waveforms enables us to go beyond the bodies from which they emanate."[19] He says that what is important is the waveform (we would say field) emitted by a body, not the body itself ("It is not the quantity of things that is important, but the effects they produce," ancient wise men used to say). It is not important who possesses the object; if we possess the waveform, then we have the presence of the actual object. Just like with the TFF.

Summing up, physical bodies communicate their own basic codes with the help of three primary factors: their nature (which transmits quality, as in the TFF), their volume (the quantity of space that they curve), and the form (the precise way they curve it). In other words, a mass creates a wave through the universe because of own substance, volume, and shape. These are the three aspects of mass to keep in mind, and they are all determined by its basic code.

The fourth element is the code itself. It is the other side of things and can, in turn, operate directly through representations of itself: symbols, images, visualizations, and much more. Consider, for example, the sound of the sea: as it is a mechanical wave, it can imprint its rhythm in our field (just like concert music), which changes the physiology of our emotions. It could, for example, relax us. If it is only a distant hum, its acoustic waves will not have any effect, but it may be possible to listen to the hum and feel relaxed: stimulated by the noise far away, we reproduce in us the frequency of the sea and receive the same benefit.

Even in different circumstances, it is possible to turn a nonmaritime background noise into a mental image of the "sound of the sea" and obtain the same result. This is what happens when we recreate mental images with relaxation and visualization techniques: the images generate frequencies that modify those of the field, thus achieving the same effects as if the imagined thing were real. This is exactly like TFF producing pharmacological results in the absence of the medicine. Eventually, we might not need things anymore if we could recreate the frequencies. How? By reproducing them artificially (TFF, smells, virtual images, or sounds) or imagining them in appropriate ways. Along different paths, the same result is reached. What matters is the effect. Nature responds to stimulations, not so much to things themselves.

This helps us understand how certain frequencies emitted with visualization manage to break the laws of physiology, when they are able to generate frequencies of food or medicines, or radiation that can kill cancer cells. There are many possible applications of virtual frequencies if we overcome the fear of encroaching on areas we think we cannot control. We prefer hunger, disease, and death to the possibility of defeating them by means that do not receive confirmation from present economic interests, and which would ultimately reveal that we live in a virtual and fake world.

An Internet Universe

We saw that in the continuum, the network of informed matter that pervades the universe—which is simultaneously everywhere—nothing moves: it manifests. The continuum behaves like a universal computer network where communication happens not by movement but by non-localization phenomena. According to our theory, bodies have a dual nature and are composed of combined matter (the mass) and informed matter (the field). The field allows the body to remain connected to the continuum, as if through invisible "umbilical cords."

Here is an example: a man living in Australia has a bad car accident; he is not hurt, but he is very frightened and fears for his life. Shaken by the trauma, the fear, the intense emotional wave, his field starts to vibrate on a certain frequency that, going through the continuum, diffuses into the great universal network. This happens, as they say, "in real time." This is a little like putting information on the Internet: it is virtually everywhere, and whoever has the connection can access it. In our case the connection is the resonance: the vibration caused by the accident can resound instantaneously with the field of the brother of that man, who is, for example, in Europe. He immediately feels that on the other side of the world something happened.

A communication without moving waves. Our usual understanding can make it difficult to accept this thesis because it is influenced by mechanistic thinking, which states that if something is in A and I find it again in B, it is obvious that it went from A to B. Let us reflect on this: no wave can be transmitted at an instantaneous speed from one side of the earth to another, unless it is included in a continuous communication like a computer network. In the universal continuum, the information does not need to travel the A-to-B path because it is *in* A as well as in B. That is the meaning of *continuum*. Communication happens not by movement but by nonlocalization phenomena. What is in A did not move but rather *manifested* in B; it is not a movement but an appearance.

To our space-time minds effort is required to access the necessary means to explore new worlds. Even the first approaches to electronic information technology were not easy. We fantasized in the 1960s about what we then defined as "electronic calculators"! In the face of novelty and change, we hide behind distrust and fear, fantasizing about the unknown, but, ironically, as soon as the unknown becomes known and apprehension subsides, we get used to the new as if it has always been obvious. We may even wonder at those who are still suspicious and reluctant to become accustomed to it.

Let us go back to our Australian man. Besides spreading in the network and transmitting to those with whom there is resonance, the emotion of the accident is imprinted in the field like a wound whose scar could influence certain future behaviors. It also may emerge one day from the unconscious as a symbolic dream or in the course of physiotherapy or regressive hypnosis. The emotional wave can also affect other things in the vicinity: trees, streams, stones, and animals. Each registers events in the field in its own way. If the leaves of surrounding trees were connected to a galvanometer at the time of the accident, they would signal alarm. The animals of a farm in the vicinity would manifest restlessness. After a while, a sensitive person would feel uneasy while traveling on that road via contact with stones near the accident that emit sadness and a sense of fear. Finally, the lady on the nearby farm might wake up suddenly from a nightmare. These are only some of the possible ways that a dramatic even can be registered. How many times do we encounter such resonances in the course of our day without our even realizing it! Or, if we do, we just blame fate or coincidences.

Every action, gesture, word, and thought goes through the continuum of pure matter. All of this information can potentially be detected by other beings, but only a fraction reaches the conscious threshold. Heaven forbid if it were otherwise! The fact that we do not perceive all the events happening around us—even if our field does—allows us to live a more serene existence, which would otherwise be tormented by exaggerated emotional feelings. Nature protects us from too much information. People with low threshold levels know this well; they are tormented by feelings and often look neurotic.

The emotional field can generate extraordinary communications, but only under certain conditions. Some people or objects are more likely to perceive emotional events that affect them; others are able to catch events that involve humanity, such as the forecast of earthquakes and catastrophes; yet others have a much higher threshold of sensitivity and perceive nothing. The perceptions of some people are blocked by a hectic life or because they are attentive only to the rational and to the information provided by their senses.

In the homogeneous and ever-present network, the events of the world are stored as the most comprehensive database. *Akashic Record* is the name that the ancient oriental traditions have given to this non-localized container of everything said, done, and thought. Nothing escapes the vast network of memory. Anyone who did or said something not quite correct in the past, convinced that "nobody can see or hear it anyway . . . so nobody will ever know" is incorrect: everything is recorded, for eternity. That includes our thoughts: none of us is without "sin."

In the continuous network, time is one. "It is wise to say that all things are one," writes Heraclitus, and he continues, "The same things are: the alive and the dead, the alert and the sleepy, the young and the old; the former change into the latter and the latter into the former. From all things to the One, and from the One to all things." Our mind cannot grasp the concept of the oneness of time, because space-time belongs in the mind that lives in a fragmentary time where any chronological category cannot escape from "before" and "after."

The continuum, the invisible texture of the universe, is the obscure cloak of Isis, the veil of Maya, of Demeter, and of all the Great Mothers. Black is the color of the Mother and her cloak, black because it is invisible and because it is the color of generation. The black sun that many people have worshiped is the feminine sun from which the world originates. Death is black; the regeneration phase of the alchemist is that of *nigredo,* or "blackness"; black plays on the stage of black opera. The black cloak of Isis is the network of pure matter spread everywhere, the obscure forge that generates things. For this is *splendescens atro nitore,* the oxymoron used by Apuleius to characterize an existence outside of space-time constraint. It is the weft, the origin of everything; the memory of everything is the *Matrix,* which supports *Nutrix* with magical powers, with its nonlocality and its miracles; *Virgo* is undifferentiated, in whose purity everything is possible and nothing acts, peace, the immaculate conception. To lift the veil is not granted to mortals. Only certain initiated ones, when ready, can go near.

Under the Veil of Isis

To lift the veil means to access the dark matter. One can then take advantage of those terrible faculties of the black cloak: the nonlocality, for example; the nontemporality, with which it is possible "to travel" in time. Some psychics can recover ancient memories from resonances at appropriate sites of the network and are even able to report on discussions and images, often verifiable. Carl Gustav Jung, when he conceived the idea of the universal unconscious, thought of a continuous and unique field that fills the universe, as in the alchemic and oriental traditions; even though, under the influence of psychoanalysis, he called it "unconscious," he certainly conceived of it as far more vast and more complex than the Freudian version.

As a doctor I have witnessed many phenomena of mediumship. For example, during her first trip to Egypt a patient of mine encountered strange visions concerning the eighteenth Egyptian dynasty. She experienced sudden visions that overlapped her real images, so much so that, in such moments, she lost touch with reality. Her traveling companions saw her making gestures into space and opening her mouth as if she wanted speak. She seemed to be in a trance. When the visions finished, she remained disoriented and tired. These altered states of consciousness were experienced as a virtual reality that involved all her senses, so precise that she was able to describe in detail the surroundings she was "transported" into. Her husband, an archaeologist and Egyptologist, has been able to verify the accuracy of her images, as he undertook excavations in places indicated by his wife, and her visions were confirmed.

It could have been hysteria, I thought at first, but I gathered evidence to the contrary. And even if she were a hysteric, this would still not explain the nature of the visions. Psychotherapists should perhaps consider certain hysterical neuroses as manifestations of sensitivity, sometimes only temporary, and should keep in mind that some hysteria can even facilitate real contacts with the network.

Among the most famous psychics in the world and capable of making "trips" in the dimension of the past was a doctor from Turin, Gustavo Adolfo Rol. He could revive any moment in history (the precise day, month, and year) in any place (physical coordinates) and recreate a virtual reality of the event, for himself and others. A person working with him might find himself a helpless spectator in the middle of the battle of Marengo, with troops of Napoleon at full charge. Many times Rol reported on items encountered during the trip that then proved authentic (such as coins of the era, or accessories from the uniforms, or weapons on the battle field). Trips to which Rol subjected his audiences even went in the opposite direction, into the future. I met a man who in the 1980s was transported by Rol in Turin to 2010. Trips to the future, even if historically not verifiable, are clues about the chronologic continuum of the network.

Allow me a brief parenthesis on the question: Are events already pre-arranged? Destiny, yes or no? The dispute between the supporters of destiny and those of free will may not make any sense, because both concepts may coexist. If life is already written, it is also true that it can be interpreted in different ways, with the script modified by free will such as, for example, reciting comically something considered a tragedy, and vice versa. Destiny may be like the commedia dell'arte, a simple "canvas" where it is written in what place we should be at a given moment and where we have to go next. Even if destiny fixes the basic points of existence, we choose the path, and we have endless possibilities available. Destiny selects the list of probable events and choices: the rest is our will.

Abilities such as Rol's are exceptional in the vast panorama of human "normality," but what is normality? Only one case is enough to destroy the limits of what we believe possible. A case alone does not "create a statistic," but offers a *possibility*. It is possible to violate certain space-time laws. It is possible that there is something beyond the senses. There are anomalies because there is another side of things. Particular people aside (who should be studied, just like Rol encouraging doctors to conduct scientific experiments on him), it is not rare that some information of the network "breaks" the space-time barriers to manifest into our consciousness. It may result in telepathic phenomenon, precognitions, or evocations of past events.

A case of replication of an event happened in France in the 1950s, affecting two English female tourists who were awakened by insistent gun fighting. An investigation confirmed that they had heard the roaring sound of a massacre that happened during the war right in that place: it was an Anglo-American raid of allied forces against the Germans, nine years earlier.[20] It seems possible to access past events in a voluntary way as well as by spontaneous resonances: the network is a library of everything that has been. Maybe one day we will able to manipulate the reality of our senses so much as to recall past images with the same ease with which we recall images from our computer.

As we have seen, genetic and emotional affinity—like between a dog and its owner, or a plant and the person who loves it—make communications easier. Among blood relations identical twins are the best communicators because they have more affinity; they shared an egg. A mother can *feel* from a distance the precise moment her child has the intention of coming home, because—via the network—the child's emotional field resonates immediately with that of the mother. There was a case reported about an English sailor who, serving during the Second World War, couldn't send letters home.[21] He had been away for more than two years, but one morning his mother went to make her son's bed and said that he would be back that evening. The family, who asked her why she was so sure about it, laughed at her. "I know and that is that," she answered when she went to prepare the boy's bedroom, and sure enough he came back that same night. Only a real intuition, not just a thought, is able to resonate as a message through the field from a child to the mother.

Everything is "video recorded" on the network, but our individual field selects only a few messages with particular significance and resonance. The emotional fields of people with close relationships are always in communication, but only emotions beyond a certain threshold can be revealed to consciousness. The fields are refined by being together, by sharing, by loving, so that, for example, after years of marriage some couples end up resembling each other, and people will say, "They look so well together. They look similar. They are made for each other." This is like what happened to the two leaves examined by Kirlian photography whose fields merged into one. Field resonance is what makes two people attracted to one another at first sight. This unexplainable magic is translated in different ways: "It is a matter of chemistry," "That woman has gotten under my skin," "He may not be handsome, but there is something attractive about him," and so on. It is normal for two people who are so in love that it seems they have fused together (we say, "two bodies and one soul") to think the same things, *feel each other* at a distance, and anticipate each another's thoughts. These are all phenomena of the field.

Our existence is full of virtual reality moments, unreported experiences that demonstrate that virtual events are possible. Why don't we think then that all of our existence, that the entire universe is just a virtual reality? That what we perceive are only shadows and it perhaps is time to leave the cave?

We pick up a lot more information from the network than we realize, because the natural threshold of defense prevents it from gaining access to our awareness. Sensitive individuals have lower defense thresholds, which allow for a greater transference of perceptions to consciousness. They can sense earthquakes, accidents, or catastrophes; or psychosomatic pains or physical symptoms can forecast, "Something is about to happen."

Precognition is knowing the occurrence of an event in advance. This is another phenomenon that can be explained by the theory of fields. Someone has premonitions because she is aware of information transferred into her field (directly from the network or from somebody else who registered the news, without being able to *read it*). Years ago when I was talking to a psychic I witnessed her suddenly turning pale. Her face became sad, and she said she was perceiving, there and then, that something terrible that was going to happen in a few days, which was going to cause me great pain. Two days later, a plane with Italian passengers crashed against a mountain of St. Mary of the Azores with no survivors. Among the victims was my friend Tatiana, a flight attendant on that plane. What did the field of the psychic resonate with? With the universal network or with my own emotional field that, unknown to me, had already grasped the information of the tragedy? There remains a mystery of something perceived before it's happening. But are we sure it is *before*?

Between the continuous and the fragmented (Plato would say between the world of truth and the shadows) there is a temporal gap that is confusing, so it may seem that we are anticipating an event, when we really are perceiving it in the exact moment when it is happening. The days that separate that instant from the moment it appears in our dimension are a shift in time. This is like the explosion of a star a hundred light years away; when we see the light of the explosion, we know it took place a hundred years ago. But if a century ago a physicist had perceived the information of the explosion at the time it was taking place, it would have appeared to be a premonition. When the senses perceive something, we are not so sure that it is happening at that place and at that time.

13 ⊶ THE WORLD OF THE STAGE MANAGERS

During our trip we have come across ordering fields in different manifestations: morphogenetic fields for plants and "vital fields" (as Harold Saxton Burr calls them) for animals. In the inorganic world, Marcel Vogel proposed ordering fields for liquid crystals and objects in general, and TFF reinforces the hypothesis of information fields in everything. Then we looked at how fields control relationships between individuals (from animals to elementary particles), with characteristics of nonlocality. And finally there are fields that direct ensembles of individuals: the cells of an organ, a flock of birds, a school of fish, and so on. Nature is organized in pyramidal hierarchies of fields informed by their own codes, which are the soul of things. Even our planet has its informed field. *Cosmos* means "beauty" but also "order," and the universe is a complex and balanced ensemble by virtue of some intrinsic system. To deny this is the case, you would have to prove—with tangible evidence—how a balanced universe could exist in the absence of controlling systems. Until then the theories presented in this book have to at least be recognized as plausible.

As we mentioned earlier, in the cellular world (the realm of organisms) the basic codes—and thus their fields—act as systems of intrinsic regulation (SIR) that direct the physiology of cellular beings; they can even predispose the being to pathologies. If something transforms the "life program" into a "cancer program," psycho-oncology suggests that profound and prolonged interior conflicts have influenced the field of the person and have thus modified the program. External factors can disturb the physiology, but they are hardly responsible for death so long as the program is life oriented.

As the SIR, the basic code is always on alert to maintain the energetic and informational homeostasis of the organism. It repairs, corrects, modifies, and chooses. It selects from the medicinal frequencies sent to the body with TFF, choosing the useful ones and discarding the others; this is why side effects are almost always absent. The informational importance of the basic codes exceeds that of the nucleic acids: the code is the essence of both organisms and objects. As the informational nucleus of the organism, it is never canceled, not even after cellular death. It is possible that this nucleus exists even before the body is created and that the matter is combined under its own attentive supervision. If this is the case, the fields generate bodies, organize them, and put them in communication with each other. This is our hypothesis.

Biological Computers

Bodies are regulated by control centers, which are made of informed matter, and the basic code acts as supervisor. An example is adrenaline flowing in the blood after a shock. Time is needed for the entire sequence of biochemical reactions as described: activation, release, and transportation of adrenaline that, subject to the speed of blood, cannot travel as fast as light. How then can we explain the fact that adrenal response to a stimulus is immediate, instantaneous? Are there other transmitters besides the molecular ones?

Since the body needs to communicate with all of its parts all the time (and in turn each is informed about everything), why don't we think about signals, instead of molecules, as information vehicles? Every being would be immersed in a field that, like a computer, controls the entire organism. In the sixteenth century Paracelsus thought that the human body was kept alive and regulated by a subtle substance he called *iliaster,* able to behave sometimes as matter and sometimes as energy. Maybe these were not Renaissance fantasies. If this control structure were molecular, there would be the need for another one to regulate it, and we should still proceed in stages up to the idea of a *nonmolecular* regulation. If bricks (molecules and cells for an organism) are the building blocks, the design must be of a different nature: an idea printed on paper.

The basic code is everywhere in the body and works with quick adjustments, regardless of molecules and electromagnetic transmissions; it is also linked to the slower molecular regulations that follow physiological and biochemical laws. Bodies are formed by molecular and cellular cohesions, but it is not clear under what principle the molecules "decide" to form in one way instead of another. Saying that the form occurs because polar forces unite molecules together does not explain the phenomenon; it only describes it. The truth is that we still do not know the design that brings the molecules together into the shape. They are kept together thanks to the forces of cohesion, but who tells them to join in that way? Bodies are not reproduced at random out of fortuitous aggregations of molecules. They are organized because something orchestrates that organization.

The Human Genome Project argues that genes absolutely control all the processes of heredity and life. How can they, if human genes number only thirty thousand, almost as many as those in a mustard plant and only twice as many genes as that of a fly or a worm? If we consider life only from a genetic point of view, a human could easily be mistaken for a mouse, whose genes are 99 percent similar.[1] We would then need to accept the theory of *alternative splicing,* according to which a single gene can encode thousands of different proteins,[2] thus contradicting the theory of Francis Crick, who codiscovered DNA. It is not believable, however, that the effect of a gene can be predicted only on the basis of its molecular sequence. In recent years the role of DNA has been redimensioned along with the dominance of the gene. All of molecular biology is faltering. The discovery that a human genome is not all that different from that of a worm pushed Eric Lander, one of the leaders of the Human Genome Project, to declare that humanity will have to learn a lesson in humility.[3] According to Barry Commoner, the director of the Critical Genetics Project of Queens College in New York, it is not the DNA molecule alone that duplicates but the entire living cell in its complexity. So here once again is something—the system itself as a whole—that could direct the operations.

Epigenetics

When Charles Darwin formulated the theory of evolution, it was not then clear how new characteristics of species could emerge or how the typical characteristics were kept through each generation. The solution came from Gregor Mendel, who postulated the "units responsible for heredity" (later called *genes*) that do not mix with each other but are transmitted intact through the generations. Ever since James Watson and Crick discovered the structure of DNA, genetic stability has been attributed to the double helix, which self-replicates, and mutations regarded as random errors. In other words, the genes have been considered as the stable units of the transmission of hereditary characteristics. Recently, however, some people think that they are not the only ones responsible for life. Difficulties arise when we ask *how* stability is maintained; then we face a "far more complex problem than was ever imagined."[4]

Let us think about duplicating chromosomes, in which the DNA chains divide into a myriad of purine and pyrimidine bases (constituents of the DNA chains, which connect the two chains, like rungs of a ladder), which absolutely must remain intact. Let us imagine these molecules like twisted filaments, so many as to seem infinite, thinner than a hair, as frail as glass, writhing around like a belly dancer while they unwind and free themselves from the helicoidal hug that kept them stable. They are trying not to break and to save all their nucleotides so that they don't remain stuck on the other side. Loss of even a fragment could cause a genetic mistake.

Once separated, each chain serves as a model for the building of its complement, and this is done with absolute fidelity: transcription errors or mutations never exceed the limit of one in ten billion. Such accuracy depends not only on the physical structure of the DNA, which alone would not even be able to replicate because it needs the help of enzymes to do so.[5] The enzymes help to prevent twisting, aid in the selection of suitable bases, control the insertion, and repair damage. But who regulates the enzymes? Who directs what happens in the DNA? Not the DNA. It is like the complicated dressing of a queen by a crowd of ladies in waiting: the queen cannot coordinate the complexity of movements around her; she cannot exercise her role because she is not even dressed. The nakedness of the queen is that of half a molecule of DNA, which is incomplete and nonoperational. Who is in control of the monitoring operations, the testing and repairing?

Genetic stability is not intrinsic to the DNA; it is an emergent property of the complex dynamics of the entire cellular network,[6] the result of a well-orchestrated process.[7] "In the conventional neo-Darwinist view, DNA is seen as an inherently stable molecule . . . and evolution, accordingly, as being driven by pure chance," Fritjof Capra writes, while instead we should "adopt the radically different view that mutations are actively generated and regulated by the cell's epigenetic network, and that evolution is an integral part of the self-organization of the living organisms."[8] The molecular biologist James Shapiro suggests thinking about rapid restructuring of the genome guided by biological feedback networks.[9]

As far as the function of genes is concerned, Francis Crick had determined that they encode the enzymes that catalyze all the cellular processes: the DNA makes the RNA, which in turn makes proteins, and proteins make us. This has been called the central dogma of molecular biology: genes determine biological traits and behavior, through a one-way flow of information from the genes to the proteins, with no feedback in the opposite direction. For the followers of genetic determinism, genes determine behavior and not the opposite. But dogmas always end up in some sort of fanaticism (I shun dogmas, and I don't tolerate fanaticisms, especially scientific ones), and sooner or later they end up collapsing. As Mae-wan Ho pointed out, the exclusive attention given to genes has obscured our vision of the organism as a whole, which is considered simply as a collection of genes subject to random mutations and selective forces in the environment over which it has no control.[10]

Sunset on Determinism

Without doubt it is DNA that determines that my eyes should be light and my hair dark (grey now); doubts arise when I pose questions about the form, structure, and functions of the body. Who decides what can grow and at what point to stop? Who establishes the limits of the body structures like the organs and limbs and the relationships between them? DNA is only a molecule, it does not have the *intelligence* to decide; it encodes proteins, but it does not teach them where to go, what to do and when. Something is missing. DNA is the executor of the design, not the director. Who regulates the DNA? The decision-making intelligence belongs to the system as a whole, which expresses itself in the basic code and in the field it informs. "The signal (or signals) determining the specific pattern in which the final transcript is to be formed . . . [comes from] the complex regulatory dynamics of the cell as a whole. . . . Unravelling the structure of such signaling pathways has become a major focus of contemporary molecular biology," says Ellen Fox Keller.[11]

Recently it has been discovered that the dynamics of the cellular network in which the genome is embedded determine what protein will be produced as well as its function.[12] The cell is able to modify the structure of a protein by altering the function. So it is cellular dynamics that ensure that different proteins appear from a single gene and that a single protein develops multiple functions. Contradicting the central dogma, we start to accept that the whole can influence the genes.

What are the cellular dynamics? Let us start by saying that genetic determinism has many problems: the cells of an organism, although they have the same genes, are different from one another. In other words, the genome is the same, but the schemes followed are different, so a muscle cell is different from a nerve cell and so on. The question is, What is that *something* that governs the differences in the expressions of genes? Genes cannot act alone; they must be activated and deactivated by certain *signals*.

To solve the problem of gene expression, in the 1960s, Francois Jacob and Jacques Monod distinguished between regulator genes and structural genes. According to the two Nobel Prize winners, the regulating mechanisms are genetic. Thus, staying within the paradigm of genetic determinism, they described biological development by using the metaphor of a "genetic program." Their proposal, arriving just as computer science was establishing itself, was received with acclaim. Subsequently, however, research has shown that the activation of the genes depends not on the genome but on the cell's epigenetic network, which includes a large number of cellular structures, particularly the chromatin. The DNA has to be considered as only part of the cellular network, which is highly nonlinear, containing multiple feedback loops, so that patterns of genetic activity continually change in response to changing circumstances.[13]

Fritjof Capra is a supporter of the new discipline of epigenetics, where biological forms and functions are the emergent properties of the entire network, an understanding that falls within the broader theory of complexity.[14] Almost a century ago, the embryologist Hans Driesch had demonstrated, through experiments done on the eggs of sea urchins, that a sea urchin could still reach full maturity even after he had destroyed many cells in the early stages of development of the embryo.[15] Recent genetics experiments have showed that the loss of single genes has very little effect on the functioning of the organism. These observations are incompatible with genetic determinism, but not with the hypothesis of codes and informed fields.[16]

In complexity theory, biological development is seen as a continuous unfolding of nonlinear systems as the embryo is formed. In mathematical terms it is as if the growth follows a path within a "basin of attraction."[17] It is accepted by many that the expression of a gene depends on the cellular context and may change if this changes. As molecular biologist Richard Strohman writes, we find that genes that are associated with particular diseases in mice have no such associations in humans. It consequently appears that mutations, even in key genes, may have effects or not, depending on the genetic context in which they are found.[18]

The emergent properties behave like an SIR superior to the DNA itself. *Epigenetics* means "above genetics" because the nature of these systems is neither genetic nor molecular. It is informed matter that operates as the basic code within the field of that body. We need to dare to go beyond the safe boundaries of the molecules, like the Argonauts who knew how to sail on dangerous and impossible seas.

Following the Path of the Argonauts

Maybe he does not know it, but my colleague Richard Gerber is an Argonaut. He writes:

> DNA contains all of the information necessary to instruct each cell how to do its particular job, how to manufacture its proteins, etc. What the DNA does not explain, however, is how these newly differentiated cells travel to their appropiate spatial locations in the developing baby's body. . . . It is highly likely that the spatial organization of the cells is ordered by a complex three-dimensional map of what the finished body is supposed to look like. This map or mold is the function of the bioenergetic field which accompanies the physical body. *This field, or etheric body is a holographic energy template that carries coded information for the spatial organization of the fetus as well as a roadmap for cellular repair in the event of damage to the developing organism.* . . . But the DNA is only an informational manual containing instructions that still must be acted upon by some intermediate actors in the cellular scheme of things. Those actors in the cellular scenario are the enzymes, the protein-bodied workers that carry out the many, everyday biochemical tasks. The enzymes catalyze specific reactions of chemicals either to create structure through molecular assemblies or to provide the electrochemical fire to run the cellular engines and ultimately keep the entire system working efficiently.[19]

The Matrix (the Great Mother) is the pure matter from which the basic codes emerge, maps of what the finished being will be, the morphogenetic fields that guide buds and embryos to become mature forms. The codes contain the designs for the entire temporal evolution of a being, its life program. The basic code contains all the images of the body, present and future, from the embryo to the completed being. We have seen that many geneticists agree that the "secret of life" does not reside in the gene sequences, and it is time to consider those systems of intrinsic regulation that control the organism. Above the gene democracy, power is in the hands of invisible dictators able to maneuver all, directors of physiology and pathology. Not the disease, but the susceptibility to the disease comes from the system; the pathogens are the last link in the chain that begins in the body. The SIR controls everything, starting with the DNA, deciding each time when and how to duplicate, what to translate and what not to, whether to suppress or to activate. Only something that possesses the overview and control of the whole ensemble can guide newborn proteins to their work sites, as in a network of integrated circuits. Are we still surprised, after witnessing the same complexity applied to computers for years? Cells and organism behave like electronic devices.

It is the SIR that sends myriads of signals to the body, to direct operations, to decide, for example, what external signals to accept and what not to, to maintain homeostasis. Perhaps the scientific community is kept away from the idea of an ordering field because nature's informed matter cannot be seen or measured. TFF started suggesting the idea of the basic codes and revealing the actions of fields; now there are instrumental confirmations, theories, and hypotheses that suggest new paradigms. There have been changes in scientific thought in recent years, such as important changes in our understanding of two major centers of control of the organism—DNA and the nervous system. With regard to the nervous system, we are moving toward the idea of a nonlocalized control center with less defined characteristics than any other anatomic structure. Let us see what this is.

Somatic Markers

The Portuguese doctor Antonio R. Damasio, professor of neurology and chairman of the Department of Neurology at the University of

Iowa College of Medicine, and professor in charge at the Salk Institute for Biological Studies in La Jolla, California, formulated the hypothesis of the "somatic marker," something nonmolecular that marks an image and can produce visceral and nonvisceral effects.

> [A somatic marker] forces attention on the negative outcome to which a given action may lead, and functions as an automatic alarm signal which says: Beware of danger ahead if you choose the option which leads to this outcome. The signal may lead you to reject, *immediately*, the negative course of action and thus make you choose among other alternatives; the automated signal protects you against future losses, without further ado, and then allows you *to choose from among fewer alternatives*. . . . *[S]omatic markers are a special instance of feelings generated from secondary emotions.* Those emotions and feelings *have been connected, by learning, to predicted future outcomes of certain scenarios.* When a negative somatic marker is juxtaposed to a particular future outcome the combination functions as an alarm bell. When a positive marker is juxtaposed instead, it becomes a beacon of incentive. . . . They do not deliberate for us. They assist the deliberation by highlighting some options (either dangerous or favorable), and eliminating them rapidly from the subsequent consideration. . . . You can think of them as devices that give a "sign."[20]

So, *signals of a nonmolecular* nature guide us in decisions and in the choice of behaviors by assisting the process of sorting; they result in an association between cognitive and emotive processes. Their purpose would be to ensure survival by reducing as much as possible unsatisfactory physical states and achieving ones that are homeostatic, functionally balanced. Where are the signals generated? Damasio thinks that they are found not just in the brain but also in the entire body. They can also act as a covert mechanism that inhibits or stimulates the tendency to act, which "would be the source of what we call intuition, the mysterious process by which we arrive at the solution of a problem *without* reasoning toward it."[21] This is another step toward the idea of something nonmolecular holding the strings to everything.

"I suspect that before and beneath the conscious hunch, there is an nonconscious process gradually formulating a prediction for the outcome of each move," Damasio writes. Something that operates through signals, the effect of a mechanism of regulation of the nervous system, which is superior to the conscious: "The idea that it is the entire organism, rather than the body alone or the brain alone that interacts with the environment, often is discounted, if it is even considered. Yet when we see, or hear, or touch or taste or smell, body proper *and* brain participate in the interaction with the environment. . . . Perceiving the environment then is not just a matter of having the brain receive direct signals from a given stimulus, the brain also receives less direct images. . . . The organism actively modifies itself so that the interfacing can take place as well as possible. The body proper is not passive."[22]

The development of interactions with the environment—the senses—takes place *somewhere* in the body. That place is the entire body. In neuroscience it could be the equivalent of DNA, a large-scale design of circuits and systems, complete with descriptions at a micro- and macrostructural level.

Damasio argues that DNA, which is governed by regulation and order, is not sufficient to explain life, and he adds that "only a part of the circuits of the brain are specified by genes."[23] The SIR cannot be located in the genome or in the neural structures. Some animal species with limited memory, reasoning, and creativity manifest complex examples of social cooperation that suggest, according to Damasio, an "ethical structure." The behavior of many animal species takes advantage of nongenetic cultural transmission.[24] This information is not in the DNA, it exists "from before." The informed field can be modified by environmental stimuli, as in the example of the migratory birds, without any genetic mutation. Like a magnetic stripe, the field stores countless quantities of information, provided it is coherent, resonant, and relevant to the system.

14 o—• A VIRTUAL WORLD

Matrix

Our journey to the other side of things is nearly at an end: it is time to assemble our ideas. Pure matter is continuous, and combined matter is discontinuous, coalesced into mass. The code and the field are both informed matter, lying between the other two, in the apparent emptiness where exchanges take place. The elementary particles exist because they exchange quanta with pure matter, which in turn generates and nourishes them. It is impossible to conceive of matter without the idea of space, or energy without time, so—Leonardo says—life is in motion. The backstage, the antiworld of our world, is the vast sea of pure matter, the fabric upon which the universe is embroidered. Combined matter is continuously generated from it to renew whatever dies: this is how the universe breathes.

The physics of the third millennium must reconsider the principles of pure matter, where every temporal effect vanishes. It is perfectly still. What is put in motion loses purity; it combines and becomes *something* in the space-time dimension. As it is concealed by the cloak of dark matter, nobody has ever seen or will ever see the face of the wearer. The Bible speaks about it under the guise of Wisdom in the book of Proverbs. It is once again the Great Mother, which is confirmed in the Roman Missal reading associated with the Feast of the Immaculate Conception (the purity of Matrix), celebrated on December 8. Let us reread this emotionally beautiful page, thinking about pure matter.

> The Lord possessed me in the beginning of his ways, before his works of old. I have been from eternity, since the beginning, before the earth was created. The abyss was not yet but I was already conceived, when the fountains of water did not yet exist, before the mountains were established, before the hills, I was brought forth. He had not yet made the earth nor the fields, or the first element of dust of the world. When He prepared the heavens, I was there, when he set a compass upon the face of the deep, when He established the clouds above, when He strengthened the underground springs, when He compassed the sea with its bounds, and set a law to the waters that they should not pass their limits, when He placed the foundations of the earth, I was with Him, forming all things, and I was daily His delight, rejoicing always before Him, rejoicing in His terrestrial world, and my delight is with the children of men (Proverbs 8:23–31).

In contrast to pure matter, there is combined matter, which is like the "harlot who gives herself to everybody and . . . makes you lose your life" (Proverbs 7). The exchange between the world and the antiworld happens in the middle kingdom, the "third kind" of Plato, the "space that offers place to the things that are generated," the threshold of the world stage. Passing through the middle, the frequencies of the backstage are translated into images, which appear as solid and real things to the senses. The key to virtual reality is to be found in the third realm.

To human eyes the real (the backstage) seems virtual, and the virtual seems real. If we could free ourselves from the senses, we would discover that things are actually solid holograms produced by the game on the virtual world stage. They exist in other forms. The mind, which in turn belongs to the virtual reality, cannot imagine forms that are different from those apprehended by the senses. These forms are the fields, the essence and true reality of things. This is the secret of all secrets, the secret humans have been chasing after for millennia. This secret has already been known and passed on by the greatest mystics, artists, philosophers, and teachers such as Plato, Jesus, Buddha, Leonardo, Giordano Bruno, Newton, and so on. The world is a virtual game in which we have agreed to represent ourselves. We continue to build it while we interact in it. The world is alchemy. What is above is similar to what is below, because everything is virtual.

Dante Alighieri, who knew about alchemy, represented the world stage as Hell, the middle kingdom as Purgatory, and the backstage as Paradise, going from black in the first, to white in the second, and red in the third. The middle kingdom, the mountain of Purgatory, complements and mirrors the infernal pit exactly in the same way as the fields belonging to this realm are virtual copies of the props. In the *Vita Nova,* Dante describes an unreal Beatrice, a symbol of pure matter (Beatrix = Matrix), "the glorious woman of mind," *Virgo* pure and uncontaminated. Virginity is a mental quality; it is Wisdom, as in Proverbs, or the virgin (*parthenos*) Athena-Minerva, who was born from the head of Zeus-Jupiter. Beatrix is dressed in red, like the Great Mother of mythology (the pre-Columbian people also portrayed *Pacha Mama* dressed in red). Red is the earth; red are the roses (sacred for Isis as for the Virgin Mary, roses are what the donkey chews to transform itself back into Lucius in the novel by Apuleius); red is the Marian month of May (sacred to the Celtic Mother, to whom the festival of Beltane was dedicated). When the Poet meets her, Beatrice is eight years and four months old. The numbers are not random. Four is matter, its multiple is pure matter: the eighth of December is the Immaculate Conception, while the eighth of September is the Virgin Mary's date of birth, under the sign of Virgo. The entirety of Dante's work can be read in this manner.

When pure matter ripples and starts to pulsate, then it is already combined. If the rhythms of matter meet with ours, we can perceive it. We do not perceive that which is still, so to us it remains obscure, invisible matter. Still, it exists.

Immense Submarine World

Many times we have compared pure matter to an immense sea (Maria is *stella maris*), undifferentiated, unchangeable, still. Bodies are formed when certain points in the sea ripple and take on characteristics of discontinuity. These forms already have the primeval characteristics of the body: they are the basic codes. Like a rock submerged in the sea, a part of which appears above the water, entities are far more extensive than the part that the senses can perceive. Each entity pulsates to the rhythm dictated by its basic code. Not all rocks rise up to the level of sense perception, so we can think of an underwater world teeming with architecture and forms of all kinds, invisible to the senses (fig. 14.1).

We have already noted that the world of the invisible is more extensive than the visible, because the perception threshold is limited. It is possible, for example, that the senses catch not the whole crest of a wave or a whole submerged rock, but only parts of it, just like when an object in the dark is illuminated by light beams from different directions. The brain uses the few illuminated fragments to reconstruct the form of the

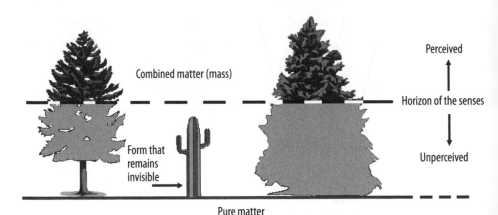

Figure 14.1. Schematic monodimensional representation of how physical bodies could be. From the flat, immobile seabed of pure matter, waves of informed matter ripple, invisible to the senses. Parts of them, which cross the threshold of perception, become mass (or combined matter). Everything that is below the threshold is the other side of things: fields, and the largest share of the mass, and forms that are invisible and inaudible to us (pure forms).

object, which is not really the object but its form "distorted" by our brain (fig. 14.2).

This is just like the effects created by Momix, the group of famous dancer-illusionists founded by the choreographer Moses Pendleton who, through games of shadows and lights, create worlds of surreal images with bodies that fly and swim, are dismantled and reassembled in visual subversions and seductions.

As rocks are anchored in the seabed, physical bodies are connected together by the continuum and the vast communication network. The physical body appearing to the senses is an interpretation of a pulsating basic code. In holography, the reality of an object is its interference pattern, which the laser beam then translates into a three-dimensional image. In the same way the true nature of things is the rhythms of the basic code that the senses translate into three-dimensional images. Things are all essentially the same. What changes is the rhythm. If the

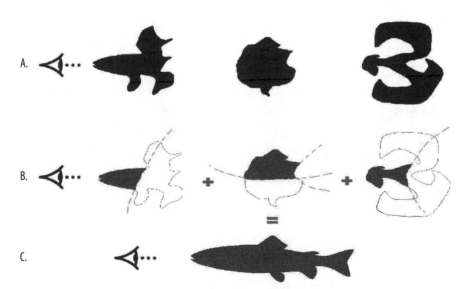

Figure 14.2. Anatomy of the invisible: which one of the three images on the top is a fish? Where is the fish? Forms are created from other bigger ones, made of informed matter, of which we perceive just a small part, which is, for us, combined matter. According to the fragments selected by our senses, the human mind reconstructs a shape that really does not exist outside of our sensory apparatus.

rhythm of iron changed to that of wood, we would perceive it as wood
and not as iron anymore.

The rhythmical pulsation of the basic codes disturbs that part of
space we call the field, and in the field, we can detect the specific frequency of the body. This is the basis of the interaction and resonance
phenomena encountered in our journey, including the TFF. Thanks to
the network, every variation of rhythm can be transmitted throughout
the world stage, with characteristics of nonlocality.

Rethinking 1687

When Newton observed the stars and imagined the planets orbiting
around the sun—Earth, for example—he was convinced that the sun
kept our planet and all the other planets in their orbits by a gravitational "leash" capable of reaching out to great distances and holding
them.[1] Einstein turned Newton's thought upside down by introducing
the idea that gravity is the curvature of space. Let us see how.

To understand this basic principle of general relativity, let us leave
time aside for a moment and imagine a flat space with only two dimensions instead of three, like a sheet of paper. This is graphically and mentally easier to represent. Let us draw a grid on the paper so it looks like
graph paper. In the absence of matter and energy, space remains flat. If
we insert a body in this space, a sphere for example (it can be a molecule
or a star), the structure of the surrounding space is deformed (fig. 14.3.)
It is like putting a bowling ball on a rubber membrane: it produces a
dent as big as the ball.

Now, let us take a very small sphere and throw it at high speed near
the large one: the ball will fall in the hole and keep rolling along the
closed curve of the depression that the bigger sphere caused in space.
It has entered into orbit. This is what the planets do around their own
sun; they follow trajectories of minimum energy in the region of distor-
tion produced by the star.

Why did we start from here? Because, with the idea of curved space,
Einstein argued that emptiness (where our journey started) is not passive

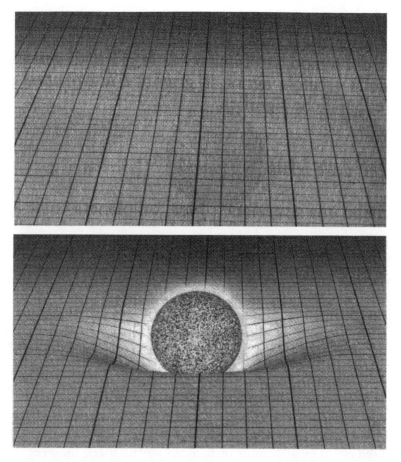

Figure 14.3. Curvature of space: above is the schematic representation of flat space; below you can see how a body (molecule or planet) curves the texture of space, like a heavy sphere on a rubber surface.

scenery, it is ether, a medium that changes depending on the objects that appear on the scene. The emptiness interacts with the solid; they are connected. Space is not just the matrix of the bodies, but is continuously shaped by them. By the mere fact of its existence, each body disturbs space at distances proportional to its mass. The agent of gravity is, according to Einstein, the pattern of the body itself. The surrounding space that is disturbed is what we call the field. The field has different components: gravitational (mass curves it), electromagnetic (from the charges), and informational (from the molecular movement).

Just like when a body is immersed in water, each object in space moves a quantity of ether equal to its volume and its form. The surrounding space is imprinted with the shape of that body. By curving space, the object makes the threads of the universal network vibrate; as a result, its presence is detectable, even at a distance, due to the resonance between fields (i.e., without using the senses). This is like when a fly is trapped in a cobweb: the threads transmit every deformation of the web itself. This is the mechanism that underlies the waveforms in radionics; entities, even if static, transmit frequencies to every point of a connected universe. "If you touch a flower, the stars vibrate," someone said. At the base is always the idea of a universal network. Laszlo writes, "There is a fundamental field that extends itself and subtly informs our own universe and all the others."[2] This is again the matrix, pure matter, the continuum that connects and informs the entire universe.

Space is always disturbed by everything, from atoms to planets. It is a question of dimension and intensity. It is evident that the perturbation of a molecule is infinitesimal to our senses, but it exists and modifies the surrounding space. Remember the "fingerprint vibrations"? They are specific for each molecule and contain information that the disturbing movement imprints in the field, making it also "informed." The intensity of emanation is influenced by the speed inherent in the matter. In TFF we excite the medicine that we want to transfer, so as to configure a prevailing interaction to enhance the transmission. If I put a perfumed essence on my hand, it will expand in the area around me, but if I shake my hand forcefully, the fragrance will be perceived more intensely and at a greater distance.

Do not think that, just because it is microscopic, the molecular perturbation of the field is negligible. It is plausible that what seems like matter to our senses is nothing more than the effect of whirling dances of particles. Let us take some examples. Let us consider a cylinder that can rotate on its longitudinal axis, and that the side facing me is solid, while the side that I do not see is empty. When the cylinder starts to rotate I will see an alternation of solid and empty, but if the rotation is increased to an inconceivably high speed, the cylinder will appear entirely solid once again. I will have the "feeling" of solid, even if it is half empty. In the same way we have a sense of a mass as having continuity and compactness, which is really only an illusion of our sensory system produced by high speeds.

Let us imagine the dance of molecules that constitutes a thing, a tennis ball, for example. Even if the field perturbation is invisible, we can imagine it as waves propagating centrifugally. If the disturbed air had a very high density (and perhaps was opaque and colored) and we looked at it through an ultramicroscope, the pulsating form of the body would finally appear to us! It is like blowing up a balloon: when it is blown up the mass of the balloon seems solid, spherical, and smooth to sight and touch; this is how the senses decode the perturbations produced by a certain vibration.

The vibration that sustains the rhythmic frequency that to the senses appears as a mass perhaps originates from the music of the vibrating strings. The rhythmic frequency we are talking about is the basic code, the sum of the information emitted by strings, particles, atoms, molecules, and so on, which is translated from this complex vibrating symphony in the mind into images and the feeling of bodies and masses. Matter is rhythm, eternal dance. There is no sound, cold, heat, color, or scent in the way we perceive it; it is really in another form. The fabric of the universe that we call space is made of ripples and hollows. Remember the hermetic text describing matter as "the crest of a wave, curling like the sea." All bodies, from atoms to stars, disturb the calm sea of pure matter, informing the universe about themselves.

A united field instantaneously connects all the parts of an organism and causes it to communicate with the environment. The organic balance needs continuous and rigorous coherence between the parts through long-range interactions: "adjustments, reactions, changes are spread in all directions at the same time."[3] The basic code is the director of everything. The genome does not contain all the sufficient instructions to build the organism. The physicist and mathematician Fred Hoyle is convinced that random construction of DNA is as probable as the assembly of an airplane by a hurricane blowing through a junkyard.[4] Without codes or designs, nature could not be expressed in its forms; neither does it proceed by trial and error or by accident; there is an order in the entire complex of changes.

Laszlo writes that there is a continuous interaction between the quantum vacuum and gravitation. Matter and emptiness react with one another in a feedback mechanism, just as in the circles of interaction between code and mass, between mass and field, and between field and code. The Hungarian author maintains that the basic radiation recorded in the whole universe represents the preexisting signals at the birth of the universe: they were already there, in its prespace.[5] The code of the entire universe, we could say, the matrix. As Böhm emphasized in the theory of the "hidden variables," the universe exists only because there are always hidden physical processes that drive the behavior of particles.

Let us go back to field perturbations. We said that the rate of molecular motion plays an important role in imprinting the rhythm in the field. This is the reason why in the transfer of a medicine it is better to excite the substance, thereby allowing it to express its rhythms more intensely. But let us ask ourselves how fast field perturbations can spread. Einstein would answer, "At the precise speed of light," so that the influence of gravity (now we are talking about planets) would keep up with the photons, without overtaking them.[6]

We know that in the microscopic world things can go differently than in the macro world. Hence the hypothesis of a network of connections that allows the perturbations to spread everywhere in real time. As we have said, it would be like a natural Internet. We talked about remote connections and instances of nonlocality. Already in 1935, Erwin Schrödinger claimed that particles do not occupy individual places, but exist in collective states in which they are intrinsically involved with each other. So particles should be considered not as individuals, but as *sociable entities* that react with each other instantly, without respecting time or space. Near or far, separated by seconds or centuries—such sociability may be due to a universal field that connects them. That this connection is intrinsic was demonstrated by

Alain Aspect's experiment. In fact spatial separation does not divide them. This is valid not just for those that have a common origin (nonlocal communications are favored by genetic and emotional links), but also for those that have originated at different points of space and time. It is sufficient that they are united with each other in the same system of coordinates.

Nonlocal connections are, as we have seen, the basis of Sheldrake's morphic field theory, of non-sensory communications with plants, animals, and human beings, and of telepathic phenomenon. We are in contact with others even when they are beyond the reach of the senses: it is enough to recall the experience described by the American doctor Marlo Morgan (described in *Mutant Message Down Under*).[7] The network allows nonlocality, and can explain parapsychological phenomena and healing operations at a distance, from radionic therapies to voodoo rituals, prayers, and healing masses and black masses, because mass is a magical ritual, as Jung taught. Distance really matters so little.

On the True Nature of Bodies

Einstein declared, "A human being is a part of the whole, called by us 'Universe,' a part limited in time and space. He experiences himself, his thoughts and feelings as something separate from the rest—a kind of optical delusion of his consciousness. This delusion is a kind of prison for us, restricting us to our personal desires, and to affection for a few persons nearest to us."

Then came the quantum and Böhm to support the existence of a field that connects things and events in a holographic way. Severi called the undisturbed quantum vacuum pure matter. Now many physicists are beginning to think that matter disturbs the quantum vacuum, that excitement and disturbance eventually create real waves in the vacuum energy sea. So pure matter becomes combined. These waves, interacting with others, form interference patterns, natural holograms that seem real to us. According to Laszlo:

Scientists know that all objects emit waves at specific frequencies that radiate outward from the objects. When the wave field emanating from one object encounters another object, a part of it is reflected from that object, and a part is absorbed by it. The object becomes energized and creates another wave field that moves back toward the object that emitted the initial wave field. The interference of the initial and the response wave fields creates an overall pattern, and this pattern carries information on the objects that created the fields. The interference pattern is effectively a hologram. The information carried in it is available at all points where the constitutive wave fields penetrate. It can be transferred from hologram to hologram if they resonate at the same frequency or at compatible frequencies.

Things in space and time are embedded in the electromagnetic field; the waves they emit are EM waves. However, things in space and time are also embedded in the cosmic plenum, and in that deeper dimension they create waves of a different kind: most likely scalar waves (these are non-vectorial waves of pure magnitude; they carry information, but not energy). The interference patterns of these waves form holograms that endure indefinitely in the plenum. The information they carry remains available for exchange with holograms resonating at compatible frequencies.[8]

It is like the sea where everything is connected with everything else through waves that mix with each other, creating interference patterns that carry information referring back to the things that produced the waves.[9] The difference is that the interference phenomena in the vacuum aren't canceled by gravity or obstacles. Laszlo says that scalar waves (discovered by Nikolas Tesla) "propagate and interfere in the vacuum and the information they transport has an effect on the particles and objects that excite the background state of the vacuum: they read and transfer information from the holograms to the vacuum."[10] He concludes by saying, "Every particle, form, object, produces perturbations that affect all other things and so everything is informed of everything, and things are directly informed by those closest to them. The particles always leave their holographic mark in the field of the vacuum."[11]

During our journey to the other side of things, we saw that "every atom sings its song continuously, and that sound creates shapes in every instant,"[12] that "forms are the result of rhythms and intermittences,"[13] and that we live by the senses and matter is an illusion created by the mind. So, what are bodies really made of? I can answer that the true nature of any thing that appears to be a body is a holographic interference pattern produced by waves similar to sound waves.

Remember that when we build an artificial hologram, the basic code of an object is contained in its interference pattern on a photographic plate; when the laser illuminates the pattern, it reconstructs a three-dimensional form, which appears real. Now imagine a similar process in nature, but the opposite in the sense that the basic codes are like interference patterns, which produce virtual objects as in the hologram. These virtual objects are the bodies that appear to our senses. Our senses perceive them as solid only because this is how they are programmed, belonging as they do to a living system that perceives itself as solid. It is the reference to ourselves that makes us look at natural holograms as solid and artificial holograms as empty. All this supports the thought that we are immersed in a virtual holographic reality.

What Is Reality?

We perceive the sensations we have inside as being outside. When we see, we are actually looking inside our brain; when we hear, we listen to a process carried out by our cortex; when we touch, we receive stimulations generated by our neurons. As in a virtual game, everything seems external to us, but really it is within us. If we trust the senses, we see everything of this virtual reality as being true. The alternative is to leave the cave and discover the deception, to realize that there is a world that our feelings do not allow us to see. How can we demonstrate that things and people exist exactly in those forms?

Embracing a crying child, I see it on my cortex; visual images are added to the acoustic and tactile sensations, and it is difficult to doubt the reality right in front of me. What I perceive, however, is not *in front* of me, but *in my nervous system*. I accept it as being in front of me because of three different simultaneous stimulations, but actually I am just experiencing more feelings. Maybe the child really is there, but not as I perceive it. Actually, what I think I perceive is not even an objective reality. So what is it?

A study on neural connections carried out by E. G. Jones and T. P. S. Powell has shown that the various cerebral ports for the sensations cannot communicate directly with each other or even with the motor areas. The cerebral structures between the areas of entry and exit are many, and the connection scheme is complex.[14] It seems that in the human brain there is no single region responsible for producing the images coming simultaneously from multiple senses. Instead, the images referring to each sensory modality are built separately.[15] Why so complex? For Damasio the complexity would serve to build, at any moment, mental images that organize information and support strategies for reasoning and decision.

The brain deceives us, making us think that inner processes are located outside the body. In the late 1960s, the Noble Prize winner, physiologist Georg von Bekesy, discovered that by placing vibrators on the knees of blindfolded people, he could induce feelings of vibration also in the space *between* the knees where there were neither vibrations nor vibration sources. There was nothing there, but the brain perceived *something* anyway, demonstrating that we have the ability to seemingly experience sensation in spatial locations where we have no sensory receptors.[16]

Another example is the phantom limb phenomenon, the sense of the presence of an amputated limb that may produce pain, itching, and cramps. For the science of medicine, the phantom limb illusion arises from nerve endings in the stump of the amputated limb or new nerve connections that are formed in the brain to redraw the areas that previously received impulses from the amputated limb. This is also a consequence of our living only by sensations: the limb is not there, but we *feel* it anyway. External reality is not that important; we live in an internal world. Sheldrake suggests that a morphic field survives the amputation and remains intact, reiterating the form of the limb. Observations of animals seem to confirm this, such as a dog refusing to lie down in the empty space left by the amputated leg of its owner.[17] The basic code is not missing and remains the guardian of the form.

The field extends beyond the physical body and can remain intact even after a part is removed. This is what indicates that our identity is the basic code, not the body. When we think about ourselves and say "I" in our mind, we imagine the body, when instead we should start to identify with the field, our subtle, invisible "part" that is the real "I." A body is only one of the possible manifestations of the basic code: the reality is a lot more complex.

An Implicit Order

While David Böhm, a leading quantum physicist and a student of Einstein and Oppenheimer, was conducting research at Berkeley, he observed that in plasma (a gas containing a high density of electrons and positive ions) electrons act not as individuals but as part of a collective. Although their individual movements appeared random, vast numbers of electrons produced effects that were well organized, as if the sea of electrons was alive, capable of isolating all impurities in a wall in the same way that a biological organism might enclose them in a cyst.[18]

Once again we see that a group behaves as a single autonomous structure, like fish schools and flocks of birds, or iron filings aligning along the lines of a magnetic force field. What is valid for organisms is also valid for things, and this appears to be a universal law: when they belong to a group, individuals no longer behave as such; they are merged in the whole. In the continuum of a flock or a school and so on, quantum effects trigger superradiance, in which the field directs each part that in turn contributes to maintain the field.

Böhm came to see the entire universe as being regulated by a complex field operating at a subquantum (quantum potential) level. When he published the theory in 1952, he had already formulated the idea of nonlocality: each particle is nonlocally connected with the others, as quantum potential pervades the entirety of space. In the physics of the fields, "where" does not exist anymore; each point of space is identical to all the others. As it pervades, the effects of the field are not reduced by distance. These sound like Pannaria's words. We are back to the continuum of pure matter, the ether, to the countless invisible threads that connect everything together.

Let us take a jar and put a rotating cylinder inside it, then fill the space between the jar and the cylinder with glycerin. A drop of ink placed in the glycerin will float motionlessly, but if the cylinder is rotated, the ink will seem to disappear. If, however, the cylinder is rotated in the opposite direction, the ink will reappear in its drop form. By observing this simple phenomenon, Böhm had the idea that the drop went back to its original form because it always retained an "order," which is not obvious, which allows it to recompose itself as a drop. In the apparent state of "disorder" of the drop diffused in the glycerine, there is a hidden or nonmanifest (or *implicate*) order, powerful enough to recreate the *explicate* order, the drop itself. This is the same phenomenon that is seen in a holographic plate that looks like a disordered configuration with no form or significance, but contains a hidden order that is expressed in the three-dimensional hologram. Böhm concluded (expressed most fully in his 1980 book titled *Wholeness and the Implicate Order*) that this order is hidden in everything and that "the universe was itself a kind of giant, flowing hologram."[19] That order is the basic code.

15 •—• CONCLUSION,
OR PERHAPS THE
BEGINNING . . .

Holographic Universe

In 1970, Karl Pribram, a neuroscientist at Stanford University, discovered that neurons of the cortex only respond to certain bands of frequencies. This reinforces the idea that the brain behaves as a holographic system and that the visual image forming in the cortex is not the object we are looking at, but a three-dimensional hologram, an artifact. If this is true, what is the original of the hologram that we perceive? What are physical bodies in reality? If the objects we see, hear, and touch become holograms inside of us, things outside of us could be the interference patterns we are projecting as three-dimensional images. Is the reality of the object called *apple* its apple appearance, or is it its interference configuration? Is the subjective reality of the forms how they are interpreted by our senses, or is it an interference configuration of incomprehensible forms?

The holographic interference configurations are similar to informed fields because they contain all the information, the basic code. The physical body is to the informed field like the hologram is to the interference configuration. The universe could be a complex form of waves that the senses transform into three-dimensional illusions: bodies. If this is so, Plato was right about reality, namely that we just see its shadows. The images of what seems real—sun, stars, flowers, living beings, and indeed ourselves in the mirror—are nothing other than holographic deformations of entities whose appearance is different and unknown. Our senses do not know the true face of reality.

"On one hand, the face of nature is so terrifying that we couldn't stand it and we would be destroyed. On the other, it is so beautiful and radiant that, like the sun, we can see it only if hidden by a white veil or as a shimmering reflection in a mirror."[1] This is how Isis talks, the Great Mother, the Matrix: "I am all that has been, that is, and that will be, and no mortal has ever lifted my veil."

We cannot lift the veil and know the true face of nature; all we can do is trust the senses and believe that an apple is so because it has form and characteristics. It would remain an apple also in its interference pattern, as a drop of ink exists even if it is dispersed in glycerin. The bodily aspect may temporarily disappear, but the basic code can at any time re-create it. The entity we call *apple* can exist in different forms; the "form made as an apple" is only one of the possibilities. The form is only the image; its essence is a broader collection of information. The senses hide a world devoid of material forms, inhabited by the basic codes and organizational patterns: the essences.

Try to think of a domain of organized frequencies. It doesn't matter how you think of it; you are imagining an object as *nonhuman* eyes could *see* it, its basic code. Near the frequency domain that looks like an apple—I apologize to apples for making them protagonists, but they are commonly taken as an example in holography, and the apple is the symbol of knowledge—there is another domain structured differently, which the eyes see in the form of a tree. And so on.

The implicate order (that looks like disorder) is not manifested, while the explicate (that seems ordered) has a form that we can experience and understand. Consequently, the first seems like an illusion, the second like reality. But it is exactly the opposite. Sensations are mental images that look real as long as we are inside of the game of the senses, but from the outside—escaping from the cave—forms appear for what they are, illusions. Each expresses a hidden whole. Laboratory instruments record an electron because, as with the drop of ink, the whole becomes an electron in that circumstance, in agreement with Heisenberg. Even the movement would be an illusion of the eternal concealing and revealing. Nature continues to hide. David Böhm believed that the universe is the *spectral representation of another parallel dimension, nonspatial and nontemporal,* an immense hologram of a kind produced by a laser beam of unconceivable intensity, in short, a virtual reality.

The idea of a universe that combines two fundamental orders, implicit and explicit, is ancient and is found in many traditions. Tibetan Buddhism speaks of "empty" and "nonempty." The Tibetan *empty* corresponds to Böhm's implicate order: the birthplace of all things, which come out from it as an endless flow. It is Lucretius's "emptiness," the "being" of Parmenides. The Tibetan *nonempty* is the objective world of forms, the "nonbeing" of Parmenides and the "doxa" of Plato. The empty is real, the nonempty illusory. There is an ancient mantra in Japanese Buddhism, *Shiki soku ze-ku, ku soku ze-shiki* ("Matter as a measure of itself is empty, empty as a measure of itself is matter.")

Influenced by perception, we cannot understand the empty. In Hinduism, the implicate order is Brahman, which is devoid of form, yet is the creative source of every visible form, the origin from which they arise and evolve. Böhm writes that the implicate order could be called spirit. Hinduism would call it consciousness. Once again it is clear that matter emerges from consciousness. Plato also affirms it in the myth of the cave. Empedocles likens the universe to a circle whose center is everywhere and circumference nowhere, while Leibniz sees the nonlocal organization of the world (apparently aware of Buddhist knowledge) and considers the universe to consist of fundamental entities—the monads—each containing a reflection of the entire universe (holographic concept). It was Leibniz who gave the keys to the integral calculus that centuries later allowed Gabor to invent the hologram. Everything returns.

All Is One

Heraclitus said, "It is wise to agree that all is one." With no stable forms, the elementary particles are transitory aspects of a hidden order, whose complexity escapes us. Once their cycle is exhausted, they are not destroyed; they go back into the implicit order from which they came. We can say with Aristotle that they go back to being particles with potential. Pannaria would say that from the world stage, they return to the backstage of pure matter, and Böhm, from the explicate to the implicate order.

So you see, in the implicate order of a *quantum* are all the aspects—wave or particle—that can manifest, depending on the interaction of the observer. It is absurd to consider the universe as divided in parts, when they are only transitory and illusory aspects of a unique reality. The particles are fantastic images of an uninterrupted holographic movement.

Different points of view produce different representations. Seen from above, this page is a sheet of rectangular paper, a large area in space, but if we examine it at a 90° angle and focus our eyes on the edge of the sheet, the page disappears and we see a thin line, like an evanescent blade. In the illusory world, we live according to points of view, none of which can describe bodies in their entirety, only parts of them. The truth cannot be divided, dissected, or categorized because it lacks material forms; we can only classify illusions, and science is limited to describing patterns of fictions. An atom is a probabilistic aspect of the vast sea that continuously exchanges atoms between stage and backstage, removing and replacing them with such speed that we cannot notice it. The universe is a single fabric whose parts are only appearance. A particle "is" only when we observe it and translate it into an image: this is the "curse" by which the senses present us with a distorted reality, but unfortunately this is the program. The truth beyond the senses is unknowable. *Deus absconditus,* "hidden God" (as Thomas Aquinas put it) is behind all forms, hidden by the illusions of Maya.

The world is therefore a construct of our brain; an interpretation of frequencies coming from other dimensions; it is virtual. It seems as if it is made of atoms, but it is reduced to fluctuating and elusive areas of probability. The physicist Nick Herbert defined it as a "radically ambiguous and ceaselessly flowing quantum soup." Whenever we try to look at it, the quantum phenomenon appears to stop and to change back into ordinary reality.[2] It is impossible to know the real in its true form, because we can receive and process it only as our abilities allow. Herbert argues that "anything we touch turns into matter," since it is the only way our senses can translate the meaning of reality. Through the senses we can never know the reality that is always hiding. "Nature loves to hide," Heraclitus wrote.

Niels Bohr remarked that if subatomic particles exist only in the presence of an observer, it makes no sense to speak of their properties before they are observed. One who disagreed was Einstein. He couldn't really accept that particles could communicate in an instantaneous way, because he thought that nothing can travel at a higher speed than light, otherwise the time barrier would be crashed and everything would be stranded in impracticable paradoxes.

If particles are not small bricks but waves, the universe is a texture of frequencies, and what looks like matter is an interference network. For us, the *real* apple is that which ripens on the tree, and the *fake* one is the hologram, but only because it feels empty to the touch, and our parameters are calibrated on solidity. Otherwise, the world would be revealed as we have never seen it before. Matter, living beings, and things are different from how we perceive them, but Nature has provided ways for us to translate reality via the senses as images of apples, people, oceans, mountains, sunsets, and so on, more enjoyable than incomprehensible holographic patterns!

The World as a Virtual Reality

In conclusion, the world is not as it appears, even if this is the only way to understand it: we *see* it as human beings. Reality is adapted to the sensory system of our species that builds its own world image. It will never be the truth. The Italian physicist Giuliana Conforto writes, "Image is the interface between form and information. . . . The field is everlasting omnipresence: it is the cause, internal and external, of every physical body."[3]

The sensations capture only parts, and they distort those to adapt them to the sensory receptors; so the senses are deceitful and different depending on the species and the individual. Nobody perceives the reality of others, as it is transformed "to suit the senses," to become *our reality*. For example, in the program that has been given to us at birth it says that contact with water gives "a sense of wet," the vicinity of fire produces a "sense of heat," and a brick wall is impenetrable. They seem absolute, but then we meet he who can walk through walls or walk on fire, she who can wound herself without bleeding or feeling pain, and then our codified system is in crisis.

With the senses we are used to doing what we want: we cancel, invent, deform. How can we trust anything? How can research be objective when it is based on the senses that are very limited and subjective? Reality is a complex system of frequencies that the neuronal networks translate into feelings and internal images. It is subjective and virtual. Hermes Trismegistus writes, "The universe is nothing but a mental creation of everything, because in reality everything is mind." And Berkeley, "Matter does not exist, it is only an idea."

The superstrings theorists understand the universe as a great symphony in which every element vibrates and plays. We can add, in complete compliance with classical chemistry, that every particle of matter has its own mode of vibration that leaves its "fingerprint" in the universe. The discovery of TFF guides us to explore a parallel world, of which the world of senses is only a mirror and projection, with different space and time.

Someone said this with simple words, not as a physicist but as a sorcerer. Don Juan explains to Carlos Castaneda that reality has an explicit order, illusory, and an implicit one, real and very powerful. He calls the first order *tonal* and the hidden one *nagual*. The world visions of a Yaqui sorcerer and of a quantum physicist such as Böhm are very similar. "We are not solid beings. We are without limits," teaches Don Juan, as if he were talking about the hologram. The tonal is the organizer of the world, the "description that we have learned to visualize and take for granted,"[4] and it composes the laws by which we perceive the world, and it is "all that we think the world is made of."

Castaneda is talking about virtual reality when he makes Don Juan say, "The *tonal* of your time calls for you to maintain that everything dealing with your feelings and thoughts takes place within yourself. The sorcerers' *tonal* says the opposite, everything is outside."[5]

While the tonal starts at birth and ends with death, the nagual never finishes. They are different times, autonomous worlds. Under the guidance of the sorcerers, Castaneda reaches the perception and finally finds the real beyond the virtual:

> It was a crucial moment, where I found myself neither in one place nor another, but in both, as an observer who accesses two scenes simultaneously. I had the incredible feeling to be able to, in that moment, go both paths. It was enough for me that I perceived them from the subject point of view.[6]

The lesson of Don Juan is that the world we think we see is only a *description* of the world. When we are convinced we are deciding, our decision has already been taken by the nagual. Deciding is nothing but submission.

Don Juan and don Genaro taught the author that for every one of us exists that double, which we mentioned several times during our journey, and that Castaneda meets. Here is the dialogue between the author and the two sorcerers.

"Is the other like the self?"

"The other is the self," answered Don Juan.

"What's the other made of?" I asked Don Juan, after minutes of indecision.

"There is no way of knowing that," he said.

"Is it real or just an illusion?"

"It's real, of course."

"Would it be possible then to say that it is made of flesh and blood?" I asked.

"No. It would not be possible," don Genaro answered.

"But if it is as real as I am . . ."

"As real as you?" Don Juan and don Genaro interjected in unison. . . .

"Obviously the double can perform acts," I said.

"Obviously!" he replied.

"But can the double act in behalf of the self?"

"It is the self, damn it!"[7]

What appears to be our double, the basic code, is our true identity. The world is virtual: it exists, but not in this form. The sorcerers teach that every feeling is illusory, and we are inside a "bubble" from the moment of birth. Is the secret of virtual reality in the bubble? "At first the bubble is open, but then it begins to close until it has sealed us in. That bubble is our perception. We live inside that bubble all of our lives. And what we witness on its round walls is our own reflection."[8]

We could open another chapter on the physics of the virtual worlds, but time and space are against us, at least for now. A pause is needed. The journey is not finished. It has just begun; we have just set foot on the other side of things, and we can already see worlds and dimensions to explore. But this will be the subject of yet another journey.

EPILOGUE

Chen Ning Yang, Nobel Prize winner for physics, writes that the proton and the neutron are sources of the nuclear field, and the electron is a source of the electromagnetic field. For us the opposite is also valid: the field is the source of particles. We mean that field informed by the basic code that governs the structure and the form of physical bodies. In organisms it is the system that regulates physiology. It results in the exchange of information that allows everything to communicate with everything.

The basic code is informed matter, very close to pure matter, belonging to a different dimension than the combined matter that appears to the senses. Its dimension is the invisible theater of transformations where everything is played before it manifests on the world stage. Even our time might be only the result of another time.

The world of the senses is the virtual reality within which science has been questioning itself from many angles. According to the theory of Glashow-Salam-Weinberg, all matter exists in doublets: the electron and its neutrino, quarks and those paired to them, and so on. Duality dominates the world stage, and it is determined according to the exclusion principle: of the two opposites, only one is selected at a time and the other excluded. This is how the world stage works, as the philosopher Anaximander of Miletus had already guessed twenty-seven centuries ago.

This is all stage illusion. In reality nothing is determined, and, as in the paradox of Schrödinger's cat, reality—true reality—is in the indeterminate, in the coexistence of opposites, in the union of the dual (*Ex duo unus*). Virtual reality is determined; true reality, however, manifests itself. On the other side of things.

ACKNOWLEDGMENTS

Firstof all, thanks to Fausto Lanfranco, who supervised my work: this book was impossible without him.

Thanks to the following team of the A. Sorti Institute of Research: Adele Molitierno, Agnese Cremaschi, Gino Rosso, Silvia Alasia, Luisa Bellando, Roberto Luttino, Alberto Celotto, Roberto Sacchi, Franco Paccagnella, Nirmala Lall, Alessandro Natella, Federico La Rocca, Emilio Citro, Francesco Aramu, Chiara Zerbinati Citro, and Daniela Mazzillo.

Thanks to my family: my parents, my wife Lea Glarey and my beautiful daughters Chiara and Gemma.

My thanks to the following allies and longtime friends: Ervin Laszlo; Fritz Albert Popp; Giuliano Preparata; Emilio Del Giudice; Claudio Cardella; Vittorio Elia; Stefania Vescia; Filippo Conti; Erich Rasche; Hans Christian Seemann; Pepe Alborghetti; Franz Morell; Pierluigi Ighina; Francesco Vignoli; Sergio Osatti; Masaru Emoto; Yasuyuki Nemoto; Santi Tofani; Christian Endler; Madeline Bastide, Roger Santini; Patrizia and Umberto Banderali; Gilles Picard; Cloe Taddei Ferretti; Gabriele Mandel; Guido Ceronetti; Franco Battiato; Marco Columbro; Niccolò Bongiorno; Giuditta Dembech; Marco Carena; Ernesto Olivero; Gabriela and Licio Gelli; Pierre Codoni; Amanda Castello; Maurizio Ghidini; Giulio Brignani; Pierluigi Bar; Claudio Gatti; Tiziana and Claudio Biglia; Roberto Romiti; Ida Domini; Alessandro Usseglio Viretta; Davide Casalini; Giorgio Papetti; Davide Boino; Riccardo Conrotto; Anna Gonella; Marina Riefolo; Eugenio

Dall'armi; Chiara Petrini; Silvia and Sebastiano Pappalardo; Giovanna De Liso; Riccardo Moffa; Franco Fusari; Alessandra Zerbinati; Valter Carasso; Giancarlo and Maia Fiorucci; Carla Perotti; Giuditta Miscioscia; Mariano Turigliatto; Rossana Becarelli; Giuseppe Lonero; Fabio De Nardis; Elio Veltri; Paolo Levi; Maria Clelia Zanini; Chiara and Lidia Ariengena; Francesca Della Valle and Gianmaria Albani; Gabriele Mieli; Ornella Gaido; Giulia Ambrosio; Mina and Bruno Zese; Laura and Zereo Chigini; Ludovica Vanni; Berenice D'Este, Francesca Tonelli, Biancarosa Romano; Katia Tonello; Rudy Lallo; Marina Lallo; Luisa Corossi Aramu; Mitsuharu Nishi; Ivan Padly; Josè Pesci-Mouttet; Léonard André; Paolo Bellavite; Margherita Nervo and Franco Boniforti; Ines Pecharroman: Paola and Pietro Bellesia; Emma Whithing and Luca Bellesia; Renzo Alberganti; Grazia and Tarcisio Zerbinati; Mariangela De Piano; Nuccia, Nando and Valeria Fantino; Margherita Montera; Adele and Michele Rosso; Marisa and Sandro Goretti; Rita and Pino Zuanazzi; Linda, Fabiola, Davide and Pietro Lapenna; Marina and Mauro Russo; Luisa Casa; Germana Frizone; Laura, Nicoletta and Maria Grazia Roncarolo; Giusi Zitoli; Gianita Bucchieri; Monica Traversa; Antonina Scolaro; Silvia Scalari and Franco Uglio; Laura Giusti; Enzo Leone; Giusi and Rosi Petraroli; Rita Volpiano; Claudia and Dario Lucchetta; Angelo and Piero Littera; Giovanna and Francesco Corso; Irma Dusio; Tiziana and Tom Bosco; Andrea Rampado; Daniela and Pier Luciano Aldrovandi; Rosy and Titti Amedeo; Manuela Pompas; Rossella De Focatis; Marco Accossato; Luca Arturi; Beppe Rosso; Valter Malosti; Carlo Bagliani; Roberto Casarin; Michele Bonetti; Enza Longo; Gino Carnazza; Patrizia Cavani; Franco Cirone; Grazia Cherchi; Clarissa Balatzeskul; Giuliana Corda; Patrizia Biancucci; Daniela and Enrico Bausano; Giorgio Ponte; Magda Cresto; Oriana and Giulio Schiavio; Silvia, Andrea and Corrado Ferroglio; Olga, Caterina and Gino Bertone; Angela, Sergio and Valter Palazzo; Gianna Chiumello; Carla, Paolo and Guido Berardo; Renato Baldassi; Adriana and Claudio Chionetti; Maria and Enzo Nuovo; Roberto Neirotti; Flavia and Antonio Toscano; Giuseppe Bormida; Peter Voss; Luisella D'Alessandro; Elisa and Daniel Keller; Claudio Villa; Gianni

Firera; Laura and Mario Gozzelino; Giacomo Passera; Simonetta, Ellison and Giovanni Carnicelli; Elena Perosino and Roberto Rorato; Anita Fico; Paola Lagorio; Ansis Abragams; Anna Benso; Raffaella Deorsola; Domenico Devoti; Fabrizio Mancin; Monica Bregola; Andrea e Regina Ospici; Alida Mazzaro; Claudia Fernandez; Sonia Rossi; Giorgio Rosso; Enzo and Giuseppe Nasillo; Tiziana Aymar; Silvia, Leo and Jacopo Giugni; Roberto Rosenthal; Mario Giacone; Mario Giaretto; Ludovica Bonanome; Giovanna Mangano; Packi Valente; Paola Ciccarelli; Paola Riva; Patrizia Brancati; Ivano Giacomelli; Pino Pelloni; Alberto Spelda; Adriana and Pietro Guglierminotti; Lucia and Mario Farina; Maria Rosa Rubatto; Amanda Castello and Paulo Parra; Raffaella Portolese; Taziana Formica; Franco Ribero; Antonello and Sergio Gentilini; Graziella Sola; Valerio Marino; Teresa Catalano; Caterina Peluso; Luisa Castellani and Paolo Masera; Maresa Rallo; Enrico Chiappini; Celeste and Domenico Molè; Fabrizio Ferragina; Antonella Eskenazi; Guido Riva; Luca Pivano; Grazia and Mario Tosi; Silvia Ferrero; Fiorella Francone; Pier Mario Biava; Alessandro Bertirotti; Teresa Totino; Grazia Monaco; Zaira Caserta; Elena Rama; Elena Ambrosin; Maria Elena Martini, Anna Zamagna; Paola Palesa; Cristina Musso; Antonello Musso; Marcello Nobili; Paolo Sacchi; Angelo, Enza, Santi, Valeria and Antonio Carlino; Gerarda, Mario, Tiziana, Paolo and Nicola Calabrese; Elisabetta Imarisio; Sabina Onomoni; Rosa Maria Sicora; Simona and Piero Grosso; Sergia and Rodolfo Luini; Narcisa Corsi; Franco Riva; Mario D'Ambrosio; Enzo Cerofolini; Patrizia Cerofolini; Gian Paolo Bucarelli; Gianluigi Mugnai; Valter Lentini, Bruna, Aldo and Paolo Paolini; Floriana Bruschi; Giotto Calbi; Alessio Basagni; Geppi and Cino Aramu; Adriana Crosetto; Adriana Terzolo; Ginevra Gheller, and all our supporters and patients whose names are in my heart and who have been helping and believing in our research.

Thanks to all who, over the years, have shared the experiments with me; the researchers and the Institutes of Research that housed and supported the experiments.

I am grateful for the English translation of Gyorgyi and Peter Byworth.

NOTES

Foreword

1. Laszlo, *Science and the Akashic Field*.
2. Laszlo, *Quantum Shift in the Global Brain: How the New Scientific Reality Can Change Us and Our World*.

Chapter 1. Prelude to Matter

1. Conforto, *Il gioco cosmico dell'Uomo*.
2. Newton, *Philosophiae naturalis principia mathematica*.
3. Severi, "Fisica subnucleare-dalla materia pura alle particelle del principio di scambio nel cronotopo" [Sub-nuclear Physics: From the Pure Matter to the Particles of the Principle of Exchange in the Chrono-topic].
4. Pannaria, *Scena e retroscena* [Scene and Back Scene]; Pannaria, "Ritorno ad Empedocle" [Back to Empedocle]; Pannaria, "Giano e la fisica" [Giano and the Physics].
5. Plato, *The Republic*.
6. Corbucci, *Alla scoperta della particella di Dio* [The Discovery of the Particle of God].
7. Gerber, *Vibrational Medicine*.
8. Ibid.
9. Burr, *Blueprint for Immortality*.
10. Ibid.
11. Ibid.

Chapter 2. The Living Vacuum

1. Talbot, *The Holographic Universe*.
2. Capra, *The Web of Life*.

3. Laszlo, *Holos: The New World of Science.*

4. Lucrezio, *De rerum natura.*

5. Campanella, *De sensu rerum.*

6. Bruno, *De rerum principiis et elementis et causis.*

7. da Vinci, *Codice Arundel.*

8. Bruno, *De rerum principiis et elementis et causis.*

9. Newton, *Philosophiae naturalis principia mathematica.*

10. Campanella, *De sensu rerum.*

11. Parmenides, *Fragment n. 12 Diehl.*

12. Muller, *Sacred Books of the East,* see "Prajñā-pāramitā-hṛdaya Sūtra."

13. Capra, *The Tao of Physics.*

14. Capek, *The Philosophical Impact of Contemporary Physics,* citing Albert Einstein, 31.

15. Preparata, *Dai quark ai cristalli* [From Quarks to Crystals].

16. Ibid.

Chapter 3. Aquatic Interlude

1. Benveniste, *Ma Vérité sur la Mémoire de l'eau* [My Truth on the Memory of Water], see preface by Brian Josephson.

2. Stillinger and Weber, "Hidden Structure in Liquids."

3. Hahnemann, *Organon dell'arte del guarire.*

4. Anagnostatos, "On the Structure of High Dilutions According to the Clatrate Model."

5. Arad, "Structure-Function Properties of Water Clusters in Proteins."

6. Lo, Lo, Chong, et al., "Anomalous States of Ice."

7. Ibid.

8. Ibid.

9. Del Giudice and Preparata, "Water as a Free Electric Dipole Laser."

10. Preparata, *Quantum Electrodynamics Coherence in Matter.*

11. Ibid.

12. Demangeat, Demangeat, Gries, Poitevin, and Costantinesco, "Modifications des temps de relaxation RNM à MHz des protons du solvant dans les très hautes dilutions salines des silice/lactose" [Changes of Relaxation Times in NMR at MHz of Proton Solvent in Very High Saline Dilutions of Silica/Lactose].

13. Ambrosini, "Indagine su differenze fisiche tra diversi campioni di H_2O con soluti ad alta diluizione" [Investigation on Physical Differences among Different Samples of H_2O with High Dilution Solutes]; Balzi, "Basi per un protocollo relativo allo studio sperimentale del rilassamento spin-spin dei

nuclei idrogeno in soluzioni acquose altamente diluite" [Basis for a Relative Protocol to the Experimental Study of the Relaxing Spin-spin of the Hydrogen Nucleus in Watery Solutions Highly Diluted]. Thesis. 1996–1997, Bologna University.

14. Elia and Niccoli, "Thermodynamics of Extremely Diluted Aqueous Solutions."

15. Ibid.; Elia and Niccoli, "New Physico-chemical Properties of Water Induced by Mechanical Treatments. A Calorimetric Study at 25°C"; Niccoli, "Proprietà termodinamiche di soluzioni ad alta diluizione" [Thermodynamic Properties of High Dilution Solutions].

16. Montagnier, Aissa, Ferris, et al. "Electromagnetic Signals Are Produced by Aqueous Nanostructures Derived from Bacterial DNA Sequences."

17. Ibid.

18. Ibid.

19. Ibid.

20. IDRAS delegation included Dr. Citro, Dr. Glarey, and Ms Molitierno (NdA).

Chapter 4. Into Nature's Nets

1. Capra, *The Web of Life*.

2. Ibid.

3. Ibid.

4. Ibid.

5. Maturana and Varela, *L'albero della conoscenza* [The Tree of Knowledge].

6. Peitigen and Richter, *The Beauty of Fractal Images of Complex Dynamical Systems*.

7. Capra, *The Web of Life*.

8. Greene, *The Elegant Universe: Superstrings, Hidden Dimensions, and the Quest for the Ultimate Theory*.

9. Ibid.

Chapter 5. Listening to Molecules Sing

1. Rilke, *Neue Gedichte*, 1907.

2. Tompkins and Bird, *The Secret Life of Plants*, 86.

3. Ibid., 87.

4. Ibid., 103.

5. Voll, *20 Jahre Elektroakupunktur*.

6. Walter, *Synopsis*.

7. Nogier, *L'homme dans l'oreille* [The Human Ear].

8. A. Sorti, personal communication, 1987.

9. Kramer, *Lehrbuch der Elektroakupunktur* [Electroacupuncture Textbook].

10. Morell, *Moratherapie: Patienteneigene und Farblicht Schwingungen, Konzept und Praxis* [Moratherapie: A Patient's Own and Colored Oscillations, Concept and Practice].

11. Fehrenbach, Noll, Nolte, et al., *Short Manual of the Vega Test-Method*.

12. Endler, Pongratz, Van Wijk, et al., "Effects of Highly Diluted Succussed Thyroxin on Metamorphosis of Highland Frog."

13. Milde, "Medikamenten Informationsübertragung."

14. Citro, "Metamolecular Informed Signal Theory and TFF."

15. J. Havel, personal communication, 2004.

16. C. Smith, International Conference, "Hidden Properties of Water" (Le proprietà nascoste dell'acqua)." The Polyclinic, Medical and Surgical Department, Napoli, 1999.

Chapter 6. The Power of TFF

1. Elia, Elia, Napoli, and Niccoli, "Condu-metric and Calorimetric Studies of the Serially Diluted and Agitated Solutions. On the Dependence of Intensive Parameters on Volume."

2. Popp, *Neue Horizonte der Medizin*.

3. Ibid.

4. Elia, Elia, Napoli, and Niccoli, "Condu-metric and Calorimetric Studies of the Serially Diluted and Agitated Solutions. On the Dependence of Intensive Parameters on Volume."

5. Ibid.

6. Citro, "TFF: un'alchimia elettronica. Basi teoriche e dati preliminari"; Citro, "TFF: dal farmaco alla frequenza."

7. Citro, "TFF: un'alchimia elettronica: Basi teoriche e dati preliminari."

8. Citro, 'TFF: dal farmaco alla frequenza."

9. Ibid.

10. Citro, "Trasferimento di Farmaci in Frequenza (TFF) e sue applicazioni nelle sindromi allergiche" [Transfer of Pharmaceutical Frequency and Its Application for Allergy Symptoms].

11. Citro, Conrotto, and Gonella, "Non Molecular Informed Signal Coming from Drugs: Possible Application in Anti-inflammatory Therapy."

12. Ibid.

13. Aissa, Litime, Attias, and Benveniste, "Molecular Signalling at High Dilution or by Means of Electronic Circuitry"; J. Benveniste, J. Aissa, M. Hjeiml, et al., "Electromagnetic Transfer of Molecular Signals." Poster at the American

Association for the Advancement of Science meeting. Boston, 1993, 1994; Citro, Penna, Papetti, and Sacchi, "L'arcano concerto che smuove una sottile energia" [The Arcane Concert That Moves a Thin Energy].

14. Citro, "Pharmacological Frequency Transfer"; G. Picard, "TFF-Glyphosate on Lentils." The First International Workshop on TFF, Torino, 1996; A. Khorassani, "TFF-Glyphosate and Trifluralin on Lentils and Wheat." The First International Workshop on TFF, Torino, 1996; M. Melelli-Roia, "Germination Tests for *Triticum aestivum* with TFF." The First International Workshop on TFF, Torino, 1996; M. Citro, et al., "2,4-D Pharmacological Frequency Transfer (TFF) on Two Different Vegetal Models." Unconventional Medicine at the Beginning of the Third Millenium, Pavia, 1998.

15. M. Citro, W. Pongratz, and P. C. Endler, "Transmission of Hormone Signal by Electronic Circuitry." Poster at the American Association for the Advancement of Science, Boston, 1993; Citro, Smith, Scott-Morley, Pongratz, and Endler, "Transfer of Information from Molecules by Means of Electronic Amplification—Preliminary Results"; Citro, Endler, Pongratz, et al., "Hormone Effects by Electronic Transmission."

16. S. Orsatti, "The TFF in Equines." The First International Workshop on TFF, Torino, 1996.

17. F. Vignoli, "Experiences in Zooiatric Practise with TFF Method." The First International Workshop on TFF, Torino, 1996.

18. Borello, *Come le pietre raccontano* [As the Stones Tell].

19. Galle, "Orientierende Untersuchung zur experimental-biologischen Überprüfung der Hypothesen zur Bioresonanz von Franz Morell" [Preliminary Study on the Experimental-Biological Bio-resonance to Verify the Hypothesis of Franz Morell].

Chapter 7. For a Science of the Invisible

1. Popp, *Neue Horizonte der Medizin*.

2. Gerber, *Vibrational Medicine*.

3. Ighina, *La scoperta dell'atomo magnetico* [Discovery of the Magnetic Atom], see www.svpvril.com/ighina/magatom.html (accessed April 8, 2011).

4. Ibid.

5. Ibid.

6. Kervran, *A la découverte des transmutations biologiques* [Discovering Biological Transmutations].

7. Ighina, *La scoperta dell'atomo magnetico* [Discovery of the Magnetic Atom], see www.svpvril.com/ighina/magatom.html (accessed April 8, 2011).

8. De Liso, "Verifica sperimentale della formazione di immagini su teli di lino

trattati con aloe e mirra in concomitanza di terremoti" [Experimental Verification of the Formation of Images on Linen Canvasses Treated with Aloe and Myrrh Concomitant of Earthquakes].

9. Forgione, *Scienza, mistica e alchimia dei cerchi nel grano.*

10. Laszlo, *Holos: The New World of Science.*

Chapter 8. Light and Music on Water

1. Mandel, *La musicoterapica dei Sufi* [The Music Therapy of the Sufi].

2. Forgione, *Scienza, mistica e alchimia dei cerchi nel grano* [Science, Mysticism, and Alchemy of Crop Circles].

3. Goethe, *La teoria dei colori* [Color Theory].

4. Forgione, *Scienza, mistica e alchimia dei cerchi nel grano* [Science, Mysticism, and Alchemy of Crop Circles].

5. Jenny, *Chimatica.*

6. Ibid.

7. Popp, *Neue Horizonte der Medizin.*

8. Forgione, *Scienza, mistica e alchimia dei cerchi nel grano* [Science, Mysticism, and Alchemy of Crop Circles].

9. Ibid.

10. Ibid.

11. David-Neel, *Tibetan Journey,* 186–87.

12. Cardella, *La Lupa e i due Soli* [La Lupa and the Two Suns].

13. *The Egyptian Coffin Texts, IV, § 261.*

14. Cardella, *La Lupa e i due Soli* [La Lupa and the Two Suns].

15. Ibid.

16. Bruno, *Opere magiche.*

17. Emoto, *The Message from Water.*

18. Forgione, *Scienza, mistica e alchimia dei cerchi nel grano* [Science, Mysticism, and Alchemy of Crop Circles].

19. Ibid.

20. Talbot, *The Holographic Universe.*

21. Lobyshev, Shikhlinskaya, and Ryzhikov, "Experimental Evidence for Intrinsic Luminescence of Water."

22. Ibid.

23. I. Bono, "Coerenza elettrodinamica nell'acqua" [Electrodynamics, Coherence of the Water]. Thesis. University of Turin, 1991–92.

24. Barber and Putterman, "Observation of Synchronous Picosecond Sonoluminescence."

25. G. Preparata, *Dai quark ai cristalli* [From Quarks to Crystals].

26. Basilio di Cesarea, *Nove omelie sull'Esamerone.*

27. Raymundo Lully, *Theatrum Chemicum,* Argentorati, 1661.

Chapter 9. Communication between Cells

1. Gurwitsch, *Das Problem der Zellteilung; Die mitogenetische Strahlung* [The Mitogenetical Radiations].

2. Reiter and Gabor, *Ultraviolette Strahlung und Zellteilung.*

3. Ibid.

4. Rajewsky, *Zehn Jahre Forschung in medizinisch-physikalischen Grenzgebiet.*

5. G. Cremonese, *Nota presentata all'Accademia pontificale delle Scienze "I nuovi Lincei,"* (Note presented to the Pontifical Academy of Science "I Nuovi Lincei"), Rome, January 21, 1920.

6. Protti, *La luce del sangue* [The Light of the Blood].

7. Ibid.

8. Ibid.

9. Protti, *L'emoinnesto intramuscolare* [The Intramuscularly Emoinnest].

10. Protti, *La luce del sangue* [The Light of the Blood].

11. Ibid.

12. Colli and Facchini, "Light Emission by Germinating Plants"; Colli, Facchini, Guidotti, Lonati, Orsenigo, and Sommaria, "Further Measurements on the Bioluminescence of the Seedlings."

13. Konev, Lyskova, and Nisenbaum, "Very Weak Bioluminescence of Cells in the Ultraviolet Region of the Spectrum and Its Biolgical Role"; Popov and Tarusov, "Nature of Spontaneous Luminescence of Animal Tissues"; Mamedov and Popov, "Ultraweak Luminescence of Various Organisms"; Veselovskii et al., "Mechanism of Ultraweak Spontaneous Luminescence of Organisms"; Zhuravlev et al., "Spontaneous Endogenous Ultraweak Luminescence of the Mitochondria of the Rat Liver in Conditions of Normal Metabolism."

14. Lawrence, "Biophysical AV data Transfer"; "Electronics and Living Plant"; "More Experiments in Electroculture."

15. W. Loos, *Naturw. Rdsch* 3 (1974): 108.

16. Kaznachev, Shurin, and Mikhailova, *Transactions of the Moscow Society of Naturalists;* Kaznachev and Mikhailova, *Sverkhslabyye Izlucheniya v Mezhkletochnykh Vzaimodeystviyakh* [Ultraweak Radiations in Intercellular Interactions].

17. Quickenden and Tilbury, "Growth Dependent Luminescence from Cultures of Normal and Respiratory Deficient *Saccharomyces Cerevisiae.*"

18. Cilento, in *Chemical and Biological Generation of Excited States.*

19. H. Klima, Dissertation. Wien: Atominstitut, 1981.

20. L. N. Kobyllyansky, *Bjull. Eksp. Biol. Med.* 94, no.11 (1982): 50.
21. Cadenas and Sies, "Low-Level Chemiluminescence of Liver Microsomal Fractions Initiated by Tertbutyl Hydroperoxide."
22. D. Bauer, Diploma work. Biophysik Ulm, 1983; Günther, "Zellstrahlungs-forschung: neue Aspekte."
23. Stempell, *Die unsichtbare Strahlung der Lebewesen* [The Invisible Rays of Organisms].
24. Galateri, *Parapsicologia ed effetto kirlian* [Parapsychology and Kirlian Effect].
25. Mandel, *Energetische Terminalpunkt-Diagnose.*
26. Presman, *Electromagnetic Fields and Life.*
27. Ostrander and Schroeder, *Scoperte psichiche dietro la cortina di ferro* [Psychic Discoveries behind the Iron Curtain].
28. Krippner and Rubin, *Galassie di vita* [Galaxies of Life].
29. Galateri, *Parapsicologia ed effetto kirlian* [Parapsychology and Kirlian Effect].
30. Mandel, *Energetische Terminalpunkt-Diagnose.*
31. Zamperini, *Energie sottili.*
32. Galateri, *Parapsicologia ed effetto kirlian* [Parapsychology and Kirlian Effect].
33. Popp, *Biologie des Lichts.*
34. Popp, *New Horizons of Medicine (The Theory of Bio Photons).*
35. Rattemeyer, Popp, and Nagl, "Evidence of Photon Emission from DNA in Living Systems."
36. Popp, *New Horizons of Medicine (The Theory of Bio Photons).*
37. Ibid.
38. Popp, Warnke, Konig, and Peschka, *Electromagnetic Bio-Information.*
39. Popp, *New Horizons of Medicine (The Theory of Bio Photons).*
40. Ibid.
41. I. Bono, *Coerenza elettrodinamica nell'acqua.* Tesi di Laurea A.A. 1991–92, Università degli Studi di Torino.
42. H. Frölich, *Advances in Electronics and Electron Physics* 5 (1980): 85.

Chapter 10. Plant and Animal Communication

1. Carlson, *The Frontiers of Science and Medicine.*
2. Tompkins and Bird, *The Secret Life of Plants.*
3. Ibid.
4. Backster, "Evidence of a Primary Perception at Cellular Level in Plant and Animal Life."

5. Tompkins and Bird, *The Secret Life of Plants.*

6. Ibid.

7. Ibid.

8. Ibid.

9. Ibid.

10. Ibid.

11. Ibid.

12. Ibid.

13. Ibid.

14. Sheldrake, *Dogs That Know When Their Owners Are Coming Home.*

15. Ibid.

16. Ibid.

17. Perdeck, "Two Types of Orientation in Migrating Starlings and Chaffinches as Revealed by Displacement Experiments."

18. Wilson, *The Social Insects.*

19. Becker, "Communication between Termites by Bio Fields."

20. Marais, *The Soul of the White Ant.*

21. Sheldrake, *Seven Experiments That Could Change The World.*

22. Hellinger, *Anerkennen, was ist Gespräche über Verstrickung und Lösung.*

23. Sheldrake, *Dogs That Know When Their Owners Are Coming Home.*

24. Ibid.

25. Ibid.

26. Castaneda, *The Teachings of Don Juan.*

27. Rosmini, *Psicologia.*

28. Rosmini, *Teosofia, vol. VI e VII.*

Chapter 11. Emotional Fields

1. Burr, *Blueprint for Immortality.*

2. Bruno, *De rerum principiis et elementis et causis.*

3. Peoc'h, "Action psychocinétique des poussins sur un générateur aléatoire" [Psychosynetic Action of Chickens by a Random Generator].

4. Simonton, Matthews-Simonton, and Creighton. *Getting Well Again;* Mambretti and Séraphin, *La medicina sottosopra. E se Hamer avesse ragione?*

5. Hasted, Böhm, et al. "Scientists Confronting the Paranormal."

6. Locke and Colligan, *The Healer Within,* 227.

7. Yogananda, *Autobiografia di uno Yogi.*

8. Rudreshananda, *Microvita: Cosmic Seeds of Life.*

9. O' Regan, "Special Report," in Tutto è uno, ed. M. Talbot, 4 (Milan: Urra, 1997).

10. Edmunds, *Hypnotism and the Supernormal*.
11. Ibid.
12. Vasiliev, *Experiments in Distant Influence*.
13. Dethlefsen, *Krankheit als Weg*.
14. Ibid.
15. Talbot, *The Holographic Universe*.
16. Ibid.
17. Ibid.
18. A. Sorti, personal communication, 1987.
19. M. Citro, "Possibile ruolo dell'immaginazione nella terapia dei tumori" [Possible Role of the Imagination in Cancer Therapy], Tesi di Spec. in Psicoterapia, Università degli Studi di Torino, A.A. 2002–2003.
20. Tofani, "Increased Mouse Survival, Tumour Growth Inhibition and Decreased Immunoreactive p53 after Exposure to Magnetic Fields."

Chapter 12. The Great Mother's Cloak of Obscurity

1. Apuleius, *L'asino d'oro* [The Golden Ass].
2. Conforto, *Il gioco cosmico dell'Uomo* [Man's Cosmic Game].
3. Bruno, *De rerum principiis et elementis et causis*.
4. Pannaria, "Giano e la fisica" [Giano and the Physics].
5. Plato, *Timeo*.
6. Ibid.
7. Burr, *Blueprint for Immortality*.
8. Sheldrake, *Dogs That Know When Their Owners Are Coming Home*.
9. Frazer, *The Golden Bough. A Study in Magic and Religion*.
10. Lakhovsky, *L'oscillazione cellulare* [The Cellular Oscillation].
11. Ibid.
12. G. Lakhovsky, *La materia* [The Matter] (Rimini: Centro di Ricerca Georges Lakhovsky, n.d.).
13. Ibid.
14. Ibid.
15. Pauli, "Moderne Beispiele zur Hintergrundsphysik" [Modern Examples of Background Physics].
16. Borello, *Come le pietre raccontano* [As the Stones Tell].
17. Ibid.
18. Einstein and Infeld, *L'evoluzione della Fisica* [The Evolution of Physics].
19. Rovere, *Chinesiologia e Naturologia*, 47.
20. Sheldrake, *Dogs That Know When Their Owners Are Coming Home*.
21. Ibid.

Chapter 13. The World of the Stage Managers

1. Capra, *The Hidden Connections: Integrating the Biological, Cognitive, and Social Dimensions of Life into a Science of Sustainability*.
2. *Science,* Feb. 2002.
3. Ibid.
4. Keller, *The Century of the Gene*.
5. Ibid.
6. Capra, *The Hidden Connections: Integrating the Biological, Cognitive, and Social Dimensions of Life into a Science of Sustainability;* Strohman, "The Coming Kuhnian Revolution in Biology."
7. Keller, *The Century of the Gene*.
8. Capra, *The Hidden Connections: Integrating the Biological, Cognitive, and Social Dimensions of Life into a Science of Sustainability*.
9. Shapiro, "Genome System Architecture and Natural Genetic Engineering in Evolution."
10. Ho, *Genetic Engineering—Dream or Nightmare?*
11. Keller, *The Century of the Gene*.
12. Ibid.
13. Capra, *The Hidden Connections: Integrating the Biological, Cognitive, and Social Dimensions of Life into a Science of Sustainability*.
14. Ibid.
15. Capra, *The Web of Life*.
16. Keller, *The Century of the Gene*.
17. Capra, *The Hidden Connections: Integrating the Biological, Cognitive, and Social Dimensions of Life into a Science of Sustainability*.
18. Strohman, "The Coming Kuhnian Revolution in Biology."
19. Gerber, *Vibrational Medicine*.
20. Damasio, *Descartes' Error: Motion, Reason, and the Human Brain*.
21. Ibid.
22. Ibid.
23. Ibid.
24. Ibid.

Chapter 14. A Virtual World

1. Greene, *The Elegant Universe*.
2. Laszlo, *Holos: The New World of Science*.
3. Ibid.
4. Ibid.
5. Ibid.

6. Greene, *The Elegant Universe.*

7. Morgan, *Mutant Message Down Under.*

8. Laszlo, *Quantum Shift in the Global Brain.*

9. Laszlo, *Holos: the New World of Science.*

10. Ibid.

11. Ibid.

12. David-Neel, *Tibetan Journey,* 186–87.

13. Jenny, *Chimatica.*

14. Talbot, *The Holographic Universe.*

15. Damasio, *Descartes' Error: Motion, Reason, and the Human Brain.*

16. Pribram, *Languages of the Brain: Experimental Paradoxes and Principles in Neuropsychology,* 169.

17. Sheldrake, *Dogs That Know When Their Owners Are Coming Home.*

18. Talbot, *The Holographic Universe.*

19. Ibid.

Chapter 15. Conclusion, or Perhaps the Beginning . . .

1. P. Citati, in *Repubblica,* February 9, 2006.

2. Herbert, "How Large is Starlight? A Brief Look at Quantum Reality."

3. Conforto, *Il gioco cosmico dell'Uomo.*

4. Castaneda, *Tales of Power.*

5. Ibid.

6. Ibid.

7. Ibid.

8. Ibid.

BIBLIOGRAPHY

Aissa, J., M. H. Litime, M. Attias, and J. Benveniste. "Molecular Signalling at High Dilution or by Means of Electronic Circuitry." *J. Immunology* 150 (1993): 146A.

Ambrosini, F. "Indagine su differenze fisiche tra diversi campioni di H_2O con soluti ad alta diluizione [Investigation on Physical Differences among Different Samples of H_2O with High Dilution Solutes]." Thesis. 1994–95, Bologna University.

Anagnostatos, G. S. "On the Structure of High Dilutions According to the Clatrate Model." In *High Dilution Effects on Cells and Integrated Systems.* Taddei-Ferretti, C., and P. Marotta, eds. London: World Scientific, 1998.

Apuleio, L. *L'asino d'oro* [The Golden Ass]. Novara: Ist. Geogr. De Agostini, 1964.

Arad, D., et al. "Structure-Function Properties of Water Clusters in Proteins." In *High Dilution Effects on Cells and Integrated Systems.* Taddei-Ferretti, C., and P. Marotta, eds. London: World Scientific, 1998.

Backster, Cleve. "Evidence of a Primary Perception at Cellular Level in Plant and Animal Life." *Internat. J. of Parapsychology* 10, no.4 (1968): 329–48.

Balzi, B. "Basi per un protocollo relativo allo studio sperimentale del rilassamento spin-spin dei nuclei idrogeno in soluzioni acquose altamente diluite" [Basis for a Relative Protocol to the Experimental Study of the Relaxing Spin-Spin of the Hydrogen Nucleus in Watery Solutions Highly Diluted]. Thesis. 1996–97, Bologna University.

Barber, B. P., and S. J. Putterman. "Observation of Synchronous Picosecond Sonoluminescence." *Nature* 352 (1991): 318–20.

Basilio di Cesarea. *Nove omelie sull'Esamerone.* Milan: Lorenzo Valla, 1990.

Becker, G. "Communication between Termites by Bio Fields." *Biological Cybernetics* 26 (1977): 41–51.

Benveniste, J. *Ma Vérité sur la Mémoire de l'eau* [My Truth on the Memory of Water]. Paris: Albin Michel, 2005.

Borello, L. *Come le pietre raccontano* [As the Stones Tell]. Cavallermaggiore: Gribaudo, 1989.

Bruno, Giordano. *De rerum principiis et elementis et causis.* Napoli: Procaccino, 1995.

———. *Opere magiche.* Milan: Adelphi, 2000.

Burr, Harold Saxton. *Blueprint for Immortality: The Electric Patterns of Life.* London: Neville Spearman, 1972.

Cadenas, E., and H. Sies. "Low-Level Chemiluminescence of Liver Microsomal Fractions Initiated by Tertbutyl Hydroperoxide." *Eur. J. Biochem.* 124 (1982): 349.

Campanella, Tommaso. *De sensu rerum.* Genova: F.lli Melita, 1987.

Capek, M. *The Philosophical Impact of Contemporary Physics.* Princeton, N.J.: D. Van Nostrand, 1961.

Capra, Fritjof. *The Hidden Connections: Integrating the Biological, Cognitive, and Social Dimensions of Life into a Science of Sustainability.* New York: Doubleday-Anchor Books, 2002.

———. *The Tao of Physics: An Exploration of the Parallels between Modern Physics and Eastern Mysticism.* Berkeley, Calif.: Shambhala Publications, 1975.

———. *The Web of Life: A New Scientific Understanding of Living Systems.* New York: Doubleday-Anchor Books, 1996.

Cardella, C. *La Lupa e i due Soli* [La Lupa and the Two Suns]. Palermo: Nuova IPSA, 1995.

Carlson, R. J. *The Frontiers of Science and Medicine.* London: Wildwood House Ltd., 1975.

Castaneda, C. *Tales of Power.* New York: Simon and Schuster, 1974.

———. *The Teachings of Don Juan.* Berkeley: University of California Press, 1968.

Cilento, G. In *Chemical and Biological Generation of Excited States.* Adam, W., and G. Cilento, eds. New York: Academic Press, 1982.

Citro, Massimo. "Metamolecular Informed Signal Theory and TFF." In *Struktur und Funktion des Wassers im Organismus.* Bergsman, O., ed., 72–77. Wien, Austria: Facultas Universitätsverlag, 1994.

Citro, M. "Pharmacological Frequency Transfer." In *High Dilution Effects on Cells and Integrated Systems.* Taddei-Ferretti, C., and P. Marotta, eds., 346–59. London: World Scientific, 1997.

———. "TFF: dal farmaco alla frequenza." *Vivibios* 2, no. 3 (1992): 66–72.

———. "TFF: un'alchimia elettronica: Basi teoriche e dati preliminari." *IPSA* 10, no. 2/3 (1992): 9–44.

———. "Trasferimento di Farmaci in Frequenza (TFF) e sue applicazioni nelle sindromi allergiche" [Transfer of Pharmaceutical Frequency and Its Application for Allergy Symptoms]. *Medicina Biologica* (1995): 45–48.

Citro, Massimo, R. Conrotto, and A. Gonella. "Non Molecular Informed Signal Coming from Drugs: Possible Application in Anti-inflammatory Therapy." VI Interscience World Conference on Inflammation, Geneva, 1995.

Citro, M., P. C. Endler, W. Pongratz, et al. "Hormone Effects by Electronic Transmission." *FASEB J.* 9 (1995): A392.

Citro, M., A. Penna, G. Papetti, and R. Sacchi. "L'arcano concerto che smuove una sottile energia" [The Arcane Concert That Moves a Thin Energy]. *Medicina Naturale* 4, no. 4 (1994): 18–23.

Citro, M., C. W. Smith, A. Scott-Morley, W. Pongratz, and P. C. Endler. "Transfer of Information from Molecules by Means of Electronic Amplification— Preliminary Results." In *Ultra-High Dilution. Physiology and Physics.* Endler, P. C., and J. Schulte, eds., 209–24. Dordrecht: Kluwer, 1994.

Colli, L., and U. Facchini. "Light Emission by Germinating Plants." *Nuovo Cimento* 12 (1954): 150.

Colli, L., U. Facchini, G. Guidotti, R. Dugnani Lonati, M. Orsenigo, and O. Sommaria. "Further Measurements on the Bioluminescence of the Seedlings." *Experientia* 11 (1955): 479–81.

Conforto, G. *Il gioco cosmico dell'Uomo.* Diegaro di Cesena: Macro, 2001.

Corbucci, Massimo. *Alla scoperta della particella di Dio* [The Discovery of the Particle of God]. Diegaro of Cesena: Macro, 2006.

Damasio, A. *Descartes' Error: Motion, Reason, and the Human Brain.* New York: Avon, 1994.

David-Neel, Alexandra. *Tibetan Journey.* London: J. Lane, 1936.

Del Giudice, E., and G. Preparata. "Water as a Free Electric Dipole Laser." *Phys. Rev. Lett.* 61 (1988): 1085–88.

De Liso, G. "Verifica sperimentale della formazione di immagini su teli di lino trattati con aloe e mirra in concomitanza di terremoti" [Experimental Verification of the Formation of Images on Linen Canvasses Treated with Aloe and Myrrh Concomitant of Earthquakes]. *Sindon N.S. Quad* 14 (2000): 125–30.

Demangeat, J. L., C. Demangeat, P. Gries, B. Poitevin, and N. Costantinesco. "Modifications des temps de relaxation RNM à MHz des protons du solvant dans les très hautes dilutions salines des silice/lactose" [Changes of Relaxation Times in NMR at MHz of Proton Solvent in Very High Saline Dilutions of Silica/Lactose]. *J. Med. Nucl. Biophy.* 16, no. 2 (1992): 135–42.

Dethlefsen, T. *Krankheit als Weg.* München: Bertelsmann Verlag, 1984.

Di Cesarea, Basilio. *Nove omelie sull'Esamerone.* Segrate, Italy: Mondadori, 1990.

Edmunds, S. *Hypnotism and the Supernormal.* London: Aquarian Press, 1967.

Einstein, A., and L. Infeld. *L'evoluzione della Fisica* [The Evolution of Physics]. Torino: Boringhieri, 1965.

Elia, V., L. Elia, E. Napoli, and M. Niccoli. "Condu-metric and Calorimetric Studies of the Serially Diluted and Agitated Solutions. On the Dependence of Intensive Parameters on Volume." *International Journal of Ecodynamics* 1, no. 4 (2006).

Elia, V., and M. Niccoli. "New Physico-chemical Properties of Water Induced by Mechanical Treatments. A Calorimetric Study at 25°C." *J. Therm. Analysis and Calorimeter* 61 (2000): 527–37.

———. "Thermodynamics of Extremely Diluted Aqueous Solutions." *Annals of Academy of Science of New York* 879 (1999): 241–48.

Emoto, Masaru. *The Message from Water.* Tokyo: HADO Kyoikusha, 2000.

Endler, P. C., W. Pongratz, R. Van Wijk, et al. "Effects of Highly Diluted Succussed Thyroxin on Metamorphosis of Highland Frogs." *Berlin. J. Res. Hom.* 1 (1991): 151–60.

Fehrenbach J., H. Noll, H. G. Nolte, et al. *Short Manual of the Vega Test-Method.* Schiltach: BER, 1986.

Forgione, A. *Scienza, mistica e alchimia dei cerchi nel grano* [Science, Mysticism, and Alchemy of Crop Circles]. Rome: Hera, 2003.

Frazer, J. G. *The Golden Bough. A Study in Magic and Religion.* New York: Oxford University Press USA, 1998.

Galateri, L. *Parapsicologia ed effetto kirlian* [Parapsychology and Kirlian Effect]. Milan: Sugar, 1978.

Galle, M. "Orientierende Untersuchung zur experimental-biologischen Überprüfung der Hypothesen zur Bioresonanz von Franz Morell" [Preliminary Study on the Experimental-Biological Bio-resonance to Verify the Hypothesis of Franz Morell]. *Erfahrungsheilkunde* 46, no. 12 (1997): 840–47.

Gerber, R. *Vibrational Medicine.* Sante Fe, N. M.: Bear & Co., 1988.

Goethe, J. W. *La teoria dei colori* [Color Theory]. Milan: Il saggiatore, 1993.

Greene, B., *The Elegant Universe: Superstrings, Hidden Dimensions and the Quest for the Ultimate Theory.* New York: W. W. Norton & Company, 1999.

Günther, K. "Zellstrahlungsforschung: neue Aspekte." *Naturwissenschaftliche Rundschau* 36, no. 10 (1983): 442.

Gurwitsch, Alexander. *Das Problem der Zellteilung.* Berlin: J. Springer, 1926.

———. *Die mitogenetische strahlung* [The Mitogenetical Radiations]. Berlin: J. Springer, 1932.

Hahnemann, C. F. S. *Organon dell'arte del guarire.* VI ed. (1842). Como: Red/ Studio Redazionale, 1985.

Hasted, J. B., D. J. Böhm, et al. "Scientists Confronting the Paranormal." *Nature* 254 (1975): 470–72.

Hellinger, B. *Anerkennen, was ist Gespräche über Verstrickung und Lösung.* München: Kösel, 1996.

Herbert, N. "How Large is Starlight? A Brief Look at Quantum Reality." *Revision* 10, no. 1 (1987): 31–35.

Ho, M-W. *Genetic Engineering—Dream or Nightmare?* Bath, U.K.: Gateway Books, 1998.

Ighina, Pier Luigi. *La scoperta dell'atomo magnetico* [Discovery of the Magnetic Atom]. Atlantide, 1999.

Jenny, H. *Chimatica*. Basel: Basilius Presse, 1974.

Kaznachev, V. P., and L. P. Mikhailova. *Sverkhslabyye Izlucheniya v Mezhkletochnykh Vzaimodeystviyakh* [Ultraweak Radiations in Intercellular Interactions]. Novosibirsk: Nauka, 1981.

Kaznachev, V. P., S. P. Shurin, and L. P. Mikhailova. *Transactions of the Moscow Society of Naturalists* 31 (1972): 224.

Keller, E. F. *The Century of the Gene*. Cambridge, Mass.: Harvard University Press, 2000.

Kervran, C. L. *A la découverte des transmutations biologiques* [Discovering Biological Transmutations]. Paris: Le courrier du livre, 1966.

Konev, S. V., T. I. Lyskova, and G. D. Nisenbaum. "Very Weak Bioluminescence of Cells in the Ultraviolet Region of the Spectrum and Its Biolgical Role." *Biophysics* 11 (1966): 410.

Kramer, F. *Lehrbuch der Elektroakupunktur* [Electroacupuncture Textbook]. Heidelberg: Haug Verlag, 1990.

Krippner, S., and D. Rubin. *Galassie di vita* [Galaxies of Life]. Torino: MEB, 1977.

Lakhovsky, G. *L'oscillazione cellulare* [The Cellular Oscillation]. Aquarius Giannone, 2010.

Laszlo, Ervin. *Holos: The New World of Science*. Milan: Riza, 2002.

———. *Quantum Shift in the Global Brain: How the New Scientific Reality Can Change Us and Our World*. Rochester, Vt.: Inner Traditions, 2008; Italian ed. Genoa: Franco Angeli, 2008.

———. *Science and the Akashic Field*. Rochester, Vt.: Inner Traditions, 2007; Italian ed. Milano: Apogeo, 2007.

Lawrence, L. G. "Biophysical AV Data Transfer." *AV Communication Review* 15, no. 2 (1972): 143–52.

———. "Electronics and Living Plant." *Electronics World* (1970): 27–29.

———. "More Experiments in Electroculture." *Popular Electronics* 93 (1971): 63–68.

Lo, S. Y., A. Lo, L. W. Chong, et al. "Anomalous States of Ice." *Modern Physics Letters B* 10, no. 19 (1996): 909–19.

Lobyshev, V. I., R. E. Shikhlinskaya, and B. D. Ryzhikov. "Experimental Evidence for Intrinsic Luminescence of Water." *J. Mol. Liquids* 82 (1999): 73–81.

Locke, S., and D. Colligan. *The Healer Within.* New York: New American Library, 1986.

Lucrezio Caro, Tim. *De rerum natura.* Florence: Sansoni, 1978.

Mambretti, G., and J. Séraphin. *La medicina sottosopra. E se Hamer avesse ragione?* Torino: Amrita, 1999.

Mamedov, T. G., and G. A. Popov. "Ultraweak Luminescence of Various Organisms." *Biophysics* 14 (1969): 1102.

Mandel, G. *La musicoterapica dei Sufi* [The Music Therapy of the Sufi]. Milan: Arcipelago, 2005.

Mandel, P. *Energetische Terminalpunkt-Diagnose.* Essen: Synthesis Verlag, 1983.

Marais, E. *The Soul of the White Ant.* Harmondsworth: Penguin, 1973.

Maturana, H., and F. Varela. *L'albero della conoscenza* [The Tree of Knowledge]. Venice: Marsilio, 1985.

Milde, K. "Medikamenten Informationsübertragung." In *Wasser,* Engler, I., ed., 121–23. Teningen: Sommer Verlag, 1989.

Montagnier, L., J. Aissa, S. Ferris, et al. "Electromagnetic Signals Are Produced by Aqueous Nanostructures Derived from Bacterial DNA Sequences." *Interdiscip. Sci. Comput. Life Sci.* 1 (2009): 81–90.

Morell, F. *Moratherapie: Patienteneigene und Farblicht Schwingungen, Konzept und Praxis* [Moratherapie: A Patient's Own and Colored Oscillations, Concept and Practice]. Heidelberg: Haug Verlag, 1987.

Morgan, M. *Mutant Message Down Under.* New York: HarperCollins, 1994.

Muller, F. M., ed. *Sacred Books of the East.* Vol. 49. New York: Oxford University Press, 1875.

Newton, Isaac. *Philosophiae naturalis principia mathematica.* Milan: Fabbri, 1996.

Niccoli, M. "Proprietà termodinamiche di soluzioni ad alta diluizione" [Thermodynamic Properties of High Dilution Solutions]. Doctorate Thesis in Research of Chemical Science. 1998–2001, Naples University "Federico II."

Nogier, P. F. *L'homme dans l'oreille* [The Human Ear]. Montreal: Maisonneuve, 1979.

Ostrander, S., and L. Schroeder. *Scoperte psichiche dietro la cortina di ferro* [Psychic Discoveries behind the Iron Curtain]. Torino: MEB, 1971.

Pannaria, Francesco. "Giano e la fisica" [Giano and the Physics]. *Civiltà delle machine* 1 (1965).

———. "Ritorno ad Empedocle" [Back to Empedocle]. *La botte e il violino* 3 (1965).

———. *Scena e retroscena* [Scene and Back Scene]. *Civiltà delle machine* 5 (1954).

Parmenides. *Fragment n. 12 Diehl.* Milan: Marcos y Marcos, 1985.

Pauli, W. "Moderne Beispiele zur Hintergrundsphysik" [Modern Examples of

Background Physics]. Reprinted in: *Wolfgang Pauli und C. G. Jung. Ein Briefwechsel,* Meier, C. A., ed. Berlin: Springer, 1992.

Peitigen, H. O., and P. H. Richter. *The Beauty of Fractal Images of Complex Dynamical Systems.* Berlin-Heidelberg-New York: Springer Verlag, 1986.

Peoc'h, R. "Action psychocinétique des poussins sur un générateur aléatoire" [Psychosynetic Action of Chickens by a Random Generator]. *Revue Française de Psychotronique* 1 (1988): 11–24.

Perdeck, A. C. "Two Types of Orientation in Migrating Starlings and Chaffinches as Revealed by Displacement Experiments." *Ardea* 46 (1958): 1–37.

Plato. *The Republic.* Florence: Sansoni, 1974.

———. *Timeo.* In *Tutte le opere.* Florence, Italy: Sansoni, 1989.

Popov, G. A., and B. N. Tarusov. "Nature of Spontaneous Luminescence of Animal Tissues." *Biophysics* 8 (1963): 372.

Popp, F. A. *Biologie des Lichts.* Berlin and Hamburg: Paul Parey Verlag, 1984.

———. *Neue Horizonte der Medizin.* Heidelberg: Haug Verlag, 1983.

———. *Nuovi orizzonti della medicina (la teoria dei biofotoni)* [New Horizons of Medicine (The Theory of Bio photons)]. Palermo: IPSA, 1986.

Popp, F. A., U. Warnke, H. L. Konig, and W. Peschka. *Electromagnetic Bio-Information.* München-Wien-Baltimore: Urban and Schwarzenberg, 1989.

Preparata, Giulano. *Dai quark ai cristalli* [From Quarks to Crystals]. Torino: Bollati Boringhieri, 2002.

Preparata, G. *Quantum Electrodynamics Coherence in Matter.* London: World Scientific, 1995.

Presman, A. S. *Electromagnetic Fields and Life.* New York: Plenum Press, 1970.

Pribram, K. *Languages of the Brain: Experimental Paradoxes and Principles in Neuropsychology.* New York: Brandon House, 1971.

Protti, Giocondo. *La luce del sangue* [The Light of the Blood]. Milan: Bompiani, 1945.

———. *L'emoinnesto intramuscolare* [The Intramuscularly Emoinnest]. Milan: Hoepli, 1932.

Quickenden, T. I., and R. N. Tilbury. "Growth Dependent Luminescence from Cultures of Normal and Respiratory Deficient *Saccharomyces Cerevisiae.*" *Photochem. Photobiol.* 37, no. 3 (1983): 337–44.

Rajewsky, B. *Zehn Jahre Forschung in medizinisch-physikalischen Grenzgebiet.* Leipzig: Hrsg. F. Dessauer, G. Thieme Verlag, 1931.

Rattemeyer, M., F. A. Popp, and W. Nagl. "Evidence of Photon Emission from DNA in Living Systems." *Naturwissenschaften* 68 (1981): 572–73.

Reiter, T., and D. Gabor. *Ultraviolette Strahlung und Zellteilung.* Berlin: Wiss. Veröffentl. A. d. Siemens-Konzern, 1928.

Rilke, R. M. *Neue Gedichte: New Poems.* Manchester, U. K.: Carcanent Press, 1992.

Rosmini, A. *Psicologia.* Rome: Città Nuova, 1981.

———. *Teosofia, vol. VI and VII.* Rome: Città Nuova, 1981.

Rovere, P. M. *Chinesiologia e Naturologia.* Rome: Marrapese, 2003.

Rudreshananda, D. *Microvita: Cosmic Seeds of Life.* Mainz: Dharma Verlag, 1988.

Severi, Francesco. "Fisica subnucleare: dalla materia pura alle particelle del principio di scambio nel cronotopo" [Sub-nuclear Physics: From Pure Matter to the Particles of the Principle of Exchange in the Chrono-topic]. *Rend. Acc. Naz.* dei XL, series IV, vol. XII, 1962.

———. "Materia e causalità. Energia e indeterminazione." *Scientia* 81 (1947): 49–59.

Shapiro, J. "Genome System Architecture and Natural Genetic Engineering in Evolution." *Annals of the New York Academy of Sciences* 870 (1999): 23–35.

Sheldrake, Rupert. *Dogs That Know When Their Owners Are Coming Home.* New York: Three Rivers Press, 2000.

———. *Seven Experiments That Could Change The World.* New York: Riverhead Trade, 1995.

Simonton, C. O., S. Matthews-Simonton, and J. L. Creighton. *Getting Well Again.* New York: Bantam, 1978.

Stempell, W. *Die unsichtbare Strahlung der Lebewesen* [The Invisible Rays of Organisms]. Jena: Fischer, 1932.

Stillinger, F. H., and T. A. Weber, "Hidden Structure in Liquids." *Phys. Rev. A* 25 (1982): 978–89.

Strohman, R. "The Coming Kuhnian Revolution in Biology." *Nature Biotechnology* 15 (1977): 194–200.

Talbot, Michael. *The Holographic Universe.* New York: Harper Perennial, 1991.

———. *Tutto è uno.* Milan: URRA, 1997.

Tofani, S., et al. "Increased Mouse Survival, Tumour Growth Inhibition and Decreased Immunoreactive p53 after Exposure to Magnetic Fields." *Bioelectromagnetics* 23, no. 3 (2002): 230–38.

Tompkins, P., and C. Bird. *The Secret Life of Plants.* London: Allen Lane, 1974.

Vasiliev, L. L. *Experiments in Distant Influence.* New York: E. P. Dutton, 1976.

Veselovskii, V. A., et al. "Mechanism of Ultraweak Spontaneous Luminescence of Organisms." *Biophysics* 8 (1963): 147.

Voll, R. *20 Jahre Elektroakupunktur.* Uelzen: Med. Liter. Verlagsgesellschaft, 1976.

Wilson, E. O. *The Social Insects.* Cambridge, Mass.: Harvard University Press, 1971.

Yogananda, Paramahansa. *Autobiografia di uno Yogi.* Rome: Astrolabio, 1971.

Zamperini, R. *Energie sottili.* Diegaro di Cesena: Macro, 1998.

Zhuravlev, A. I., et al. "Spontaneous Endogenous Ultraweak Luminescence of the Mitochondria of the Rat Liver in Conditions of Normal Metabolism." *Biophysics* 18 (1973): 1101.

INDEX

Page numbers in *italics* refer to illustrations.

ABOUT THE AUTHOR

Massimo Citro, M.D., is a physician, writer, researcher, and the discoverer of pharmacological frequency transfer (TFF). In addition to his degree in medicine, Dr. Citro took on the additional specialization of psychotherapy, studying with Lino Graziano Grandi, director of the "Alfred Adler" Institute in Turin, and completing his studies with honors and the publication rights for his dissertation. Dr. Citro also has a degree in literature, *Lettere Classiche* (Literae Humaniores), historical studies, from Turin University, with a dissertation on Jung and Gnosticism.

Dr. Citro is the founder and director of the Alberto Sorti Research Institute (IDRAS), and on the recommendation of the New York Academy of Science, he became a member of the Club of Budapest, founded by the philosopher of science Ervin Laszlo. He is the winner of the Superga Award for Literature in 2009 and the Premio Creatività e Sviluppo (Creativity and Development Award) in 2010.

Dr. Citro also has certification in acupuncture and traditional Chinese medicine, attending schools in Italy and Germany and specializing in electroacupuncture. He was a student of the great electroacupuncturist Dr. Hans Christian Seemann at Aalen, Germany. He also studied with Prof. Bruno Bruni, head of the Department of Endocrinology in Turin; Dr. Alberto Sorti in Bergamo, for metamolecular medicine and quantum physics; Adele Rosso, the vice president of European Herbalists of "Maison Verte"; Prof. Claudio Cardella of

"La Sapienza" University, Rome, with whom he studied Severi's and Pannaria's physics and the fundamentals of alchemy; and Prof. Gabriele Mandel for Sufism. His spiritual father was the Director of Centro Studi Rosminiani, Prof. Don Remo Bessero Belti, the last private confessor at Quirinale of the President of the Italian Republic at Quirinale, MP Francesco Cossiga.

A major turning point for Dr. Citro occurred in 1990 when he discovered TFF (Transfer Pharmacological Frequency), which enables the transfer of the medicinal properties of many drugs through electromagnetic circuits without administering them, thus avoiding toxicity. Far from being limited to pharmacology, it is an authentic revolution that involves chemistry, physics, and biology and paves the way for new scientific and philosophical thinking about the origin and true nature of the universe. To conduct his research on TFF, Dr. Citro founded the Alberto Sorti Research Institute in Turin (IDRAS), which he now leads.

As a researcher, Massimo Citro worked with Prof. Fritz-Albert Popp (Director of the Biophysical Institute of Kaiserslautern, Germany), Dr. Jacques Benveniste (Director of INSERM, Paris), Christian Endler, Prof. Madeleine Bastide (Department of Pharmacology, Montpellier University, France), Dr. Roger Santini (INSA, Lyon, France), Dr. Cloe Taddei-Ferretti (National Research Institute at Arco Felice, Naples), Prof. Vittorio Elia (Chemistry Institute of "Federico II" University, Naples), and Masaru Emoto (Director of the HADO Centre, Tokyo) and collaborated with Prof. Giuliano Preparata and Prof. Emilio Del Giudice, theoretical physicists, at Milan University. Received by Nobel Laureate Luc Montagnier in Paris, he is now collaborating with him on common research projects.

Dr. Citro works as a freelance practitioner in his consulting room in Turin applying psychotherapy, natural medicine, and food intolerance tests. He lives in Turin, Italy.